SSD for R

SSD for R

An R Package for Analyzing Single-Subject Data

Second Edition

CHARLES AUERBACH AND WENDY ZEITLIN

Oxford University Press is a department of the University of Oxford. It furthers
the University's objective of excellence in research, scholarship, and education
by publishing worldwide. Oxford is a registered trade mark of Oxford University
Press in the UK and certain other countries.

Published in the United States of America by Oxford University Press
198 Madison Avenue, New York, NY 10016, United States of America.

Library of Congress Cataloging-in-Publication Data
Names: Auerbach, Charles, editor. | Zeitlin, Wendy, editor.
Title: SSD for R : an R package for analyzing single-subject data /
Charles Auerbach & Wendy Zeitlin.
Description: Second edition. | New York, NY : Oxford University Press, [2022] |
Includes bibliographical references and index.
Identifiers: LCCN 2021025228 (print) | LCCN 2021025229 (ebook) |
ISBN 9780197582756 (paperback) | ISBN 9780197582770 (epub) |
ISBN 9780197582787
Subjects: LCSH: Social sciences—Research. | Single subject research.
Classification: LCC H62 .A8493 2022 (print) | LCC H62 (ebook) |
DDC 001.4/2—dc23
LC record available at https://lccn.loc.gov/2021025228
LC ebook record available at https://lccn.loc.gov/2021025229

DOI: 10.1093/oso/9780197582756.001.0001

1 3 5 7 9 8 6 4 2

Printed by Marquis, Canada

Contents

Contents

Introduction

Single-Subject Research Designs in the Behavioral and Health Sciences

Introduction

This introduction provides background information on single-subject research and its use in the behavioral and health sciences. Here you will find a brief history of the use of this type of research design to provide a backdrop for its current use and a discussion of future directions in the use of these designs. The purpose of this is to provide contextual information for the introduction of *SSD for R*, a visual and statistical software package that is easily accessible and useful in the analysis of single-subject data. *SSD for R* is a package written in R, a free and open-source statistical programming language (The R Project for Statistical Computing, n.d.). In this Introduction, we also go over the contents of this book, including updates in the second edition.

Description and Usage of Single-Subject Research

Single-subject research is also referred to in the literature as "$n = 1$ research," "interrupted time series research," and "single-case research." By whatever name, single-subject research designs are substantively different from the more commonly published group research designs; instead of examining aggregate data for multiple research subjects simultaneously, single-subject research is concerned with the empirical examination of a single research subject over time. While a single research subject could mean an individual, it could be any subject that could be conceptualized to be a single unit. For example, single-subject research could just as easily use a couple, family, therapeutic group, community, or even a larger population as its unit of analysis.

Single-subject research has some unique characteristics that differentiate it from other types of research designs. First, the commonest use of this is in the evaluation of interventions (Smith, 2012). In single-subject research, however, the subject serves as his or her own control or comparison. To accommodate this, single-subject research designs typically collect data using frequent repeated measures over time, with collected data being assigned to phases (e.g., baseline, intervention) with the baseline, or preintevention, phase often used as the control or comparison.

SSD for R. Charles Auerbach and Wendy Zeitlin, Oxford University Press. © Oxford University Press 2022.
DOI: 10.1093/oso/9780197582756.003.0001

Therefore, single-subject research is a type of time series in which the series may be interrupted with the introduction or withdrawal of one or more interventions.

Single-subject research, although quantitative, has some of the characteristics of narrative case studies that focus on individuals. Like case studies, single-subject research often contains a detailed description of the subject and, unlike group designs, also contains a detailed description of the intervention or experimental condition (Kazdin, 2011). However, with its inclusion of repeated measures, single-subject research adds a level of methodological rigor not found in case studies. To this end, single-subject research falls into the realm of scientific, quasi-experimental research, while narrative case studies are considered prescientific (Miller, n.d.).

Unlike other designs, single-subject research can have two distinct, but equally important, purposes: for formal research such as studies published in peer-reviewed journals and the less formal, but equally important, evaluation of practice. In terms of formal research, single-subject studies have been published in a wide variety of social science, health, and allied health fields, including social work, psychology, education, speech therapy, rehabilitation medicine, occupational therapy, physical therapy, and physiology. The advantages to utilizing single-subject research designs include the ability to examine conditions that are rare or to examine unique client/ patient populations (Janosky, Leininger, Hoerger, & Libkuman, 2009). They can also be used to introduce a new intervention or practice modality into the professional literature.

With regard to practice evaluation, single-subject research techniques can be used to empirically evaluate client progress over time, simply and dynamically. Techniques used to capture and evaluate client data in order to do this type of analysis are easily learned, inexpensive, and simple to implement and can be built into typical practice. The benefit of including this type of evaluation into ordinary practice is the ability to more accurately and empirically evaluate client progress longitudinally, which can then be used to inform the practitioner's work with individual client systems. That is, practitioners can readily evaluate their clients' progress over time to adapt their work to meet the specific needs of those they serve.

The History of Single-Subject Research

Single-subject research has long been part of the repertoire of inquiry in the behavioral and health sciences. Some of the earliest known research using these designs were the physiological studies conducted by Mueller and Bernard in the 1830s (Krishef, 1991). From the late 1880s through the early 1900s, a good deal of experimental psychology focused on only one or a few subjects in a given study. These included Pavlov's quantitative analyses, Breuer's qualitative cases studies, and Ebbinghaus's memory studies (Kazdin, 2011; Krishef, 1991).

In the early to mid-1900s, there was a general shift in research to studying groups instead of individuals. This has been attributed to the development and popularity of group statistical techniques such as the *t* test. Group analysis, then, gained favor as it is difficult to attain statistical significance with small sample sizes, including the typical number of small observations found in most single-subject research designs. Therefore, experimental group designs became more popular due to their scientific rigor (Kazdin, 2011; Krishef, 1991).

Learning and behavioral analysis, such as the type studied by Skinner, became the forerunner of modern single-subject research designs (Kazdin, 2011). In those studies, research was used to develop and refine theory (J. D. Smith, 2012). Additionally, many measures were behaviorally based, often noting the frequency of a target behavior. This allowed for the close examination of the impact on an individual of an intervention that simply could not be detected using group analysis. This research focused less on statistical analysis and more on observed change over time (Kazdin, 2011).

In the early 1970s, Bergin and Strupp noted the inherent limitations of group research designs. These included the high costs associated with conducting group research, the ethical issues related to assigning subjects to control conditions, the difficulty of applying results from group analyses to individual clients, and the difficulty in explaining why some subjects would improve with treatment while others did not (Krishef, 1991). More recently, questions have continued to be posed about the utility of group findings from randomized control trials to actual practice settings. This concern has resulted in a renewed and growing interest in the area of single-subject research designs (J. D. Smith, 2012).

Single-Subject Research Today

In recent years, the scientific community has seen renewed interest in single-subject research, and it is also increasing in relevance due to professional organizations that espouse high practice standards of its membership. For example, in 2008 and again in 2015, the Council on Social Work Education (CSWE), the accrediting body of schools of social work in the United States, specifically identified two areas in which single-subject research would be appropriate in its *Educational Policy and Accreditation Standards*. Competency 4 states that accredited professional social work programs should contain curriculum that builds student competency in "engag[ing] in research-informed practice and practice informed research." Additionally, Competency 9 directs programs to teach social worker students to "evaluate practice with individuals, families, groups, organizations, and communities" (CSWE, 2015). Similarly, the National Association of Social Workers' (NASW) *Code of Ethics* requires social workers to evaluate their own practices under Ethical Standard 5.02—Evaluation and Research. Specifically, "social workers should monitor and evaluate policies, the implementation of programs, and practice interventions," and "social workers should critically examine and keep current with emerging knowledge relevant to social work and fully use evaluation and research evidence in their professional practice" (NASW, 2017). While single-subject research is not the only way to evaluate clinical practice, the techniques available in these designs are easy to learn and congruent with the goals set forth by both CSWE and NASW.

Additionally, the evidence-based practice movement has attempted to bring more rigor into practice settings by incorporating scientific research into clinical work with the hope of improved service to clients (Thyer & Myers, 2011). To this end, single-subject research designs have been used to build evidence to the effective treatment of problems such as attention deficit hyperactivity disorder and other behavioral disorders, schizophrenia, and depersonalization disorder (Nathan & Gorman, 2002; Schudrich, 2012). One of the advantages of doing this type of research is the ability to evaluate interventions as they are delivered in practice settings with actual clients, which differs dramatically from group intervention designs (Beddoe, 2011). In this way, practice-based research can be used to help bridge the translational research gap between group research designs and actual practice.

Single-subject research designs are considered for inclusion in systematic reviews by respected groups such as the Cochrane and Campbell Collaborations, provided they are scientifically rigorous (Higgins & Thomas, 2020; Schlosser & Wendt, 2008; Scruggs & Mastropieri, 2013; Wendt, 2009). With systematic replication of single-subject research, studies can uncover causal relationships between an intervention and the dependent variable (Schlosser & Wendt, 2008).

In order to help elevate what constitutes scientifically rigorous single-subject studies, various groups valuing this type of research have developed research standards. These groups include the What Works Clearinghouse from the U.S. Department of Education, the National Institute of Child Health and Human Development, and several divisions of the American Psychological Association: Division 12 (clinical psychology), Division 53 (clinical child and adolescent psychology), and Division 16 (evidence-based practice in school psychology) (Smith, 2012; Thyer & Myers, 2011). Common standards found across some of these groups include

- Use of a minimum number of data points per phase (e.g., 3 to 5 in the baseline)
- More than one judge to visually analyze data to increase interrater reliability
- The use of validated measurement instruments
- The use of multiple independent measures
- The use of multiple single-subject research studies
- The requirement for alternating treatment designs (e.g., ABAB)
- Reaching levels of statistical significance between the treatment and test condition (Chambless et al., 1998; Kratochwill et al., 2013; Vannest, Davis, & Parker, 2013)

The Future of Single-Subject Research

Researchers and others interested in single-subject research designs have been increasingly productive in better understanding existing statistical methods and developing new techniques. For example, in 2014, a group of psychologists, special educators, and statisticians collaborated to introduce new methodologies and present reactions to these from leaders in the behavioral and social sciences (Kratochwill & Levin, 2014).

More recently, *Evidence-Based Communication Assessment and Intervention* published a special issue, *Advances in Statistical Analysis and Meta-analysis of Single-Case Experimental Designs* (Wendt & Rindskopf, 2020). This volume highlighted newer statistical techniques that allow for aggregating multiple single-subject studies into one, which is more rigorous than analyzing one case at a time, as well as newer measures of effect size, which are all discussed further in this text.

In our work developing *SSD for R*, we have taken many of these advances into consideration, and this second edition describes new functions that were not available when we published the original text.

Visual Analysis? Statistical Analysis?

Historically, visual analysis has been the preferred mode of examining single-subject research data, and the depiction of graphs showing data across phases has been one of the hallmarks of published single-subject research studies. For the most part, this preference for visual interpretation of data continues to exist; however, there are situations in which statistical analysis may be used to supplement visual analysis. For example, statistical analysis is appropriate in cases when treatment "effects are small; the variation is large; the effects of the intervention do not appear immediately as phases change; [or] there is no clear trend in the data within phases" (Auerbach & Schudrich, 2013, p. 346).

There are two additional concerns that have been raised with regard to solely using visual analysis to interpret single-subject research data. First, autocorrelation, or serial dependence, can greatly impact visual interpretation, and there is no way to detect autocorrelation without statistical analysis (Auerbach & Schudrich, 2013; Smith, 2012). The other issue has to do with the mechanics of visual analysis. Studies have shown that there is inconsistency in how graphs are interpreted, even when skilled raters have received the same training (Janosky et al., 2009; Kazdin, 2011; Nourbakhsh & Ottenbacher, 1994). Therefore, there can be multiple interpretations for the same data when visual analysis is the only tool that is used.

While there has been consensus that there is a need for statistical analysis in at least some single-subject research studies, there is an ongoing discussion of how this should be done. That is, it is not always clear which statistical techniques should be applied to single-subject data. However, as in group research designs, it is important to consider that the types of statistical treatments applied to single-subject research designs should be based on the characteristics of the data (J. D. Smith, 2012).

Throughout this book, then, you will notice that we often present methods to analyze data both visually and statistically, and we provide examples of how you would interpret and present your findings based on both visual and statistical analysis. These examples are applicable whether you are using these techniques for the purposes of evaluating your own practice or for the sake of producing publishable research findings.

Tools for Evaluating Single-Subject Research Data

While single-subject research has existed for many years, there has been a dearth of software to help analyze this type of data. This has been problematic because statistical software that has been intended for group research designs is often not appropriate for analyzing single-subject data.

In terms of visual analysis, it appears as if many researchers use Microsoft Excel or similar spreadsheet packages to draw line graphs. J. G. Orme and Combs-Orme (2011), for example, published a text that uses Excel macros to produce graphs specifically for the visual interpretation of single-subject data.

Another group of researchers has improvised a way of statistically analyzing visually presented data by using several software packages consecutively. First, graphs are scanned into *Ungraph* to digitize *x-y* coordinates, then *SPSS* is used to prepare the data for import and eventual analysis in *HLM* (Nagler, Rindskopf, & Shadish, 2008). While this is an interesting proposition, a couple of obvious issues arise. First, researchers have to have access to all three propriety software packages, which can be expensive. Additionally, the analytical capabilities of this combination, as described by Nagler and colleagues, are limited.

Simulation Modeling Analysis was designed for analyzing single-subject data in the healthcare arena (Borckardt, 2008). While this has been a useful tool, it also has limited visual and statistical functionality.

Pustejovsky and Swan (2018) have created a web-based effect size calculator that can be accessed through their website or from the Comprehensive R Archive Network under the package *SingleCaseES*. Their software enables users to enter data for two phases and then compute a number of effect sizes with additional output and graphs.

Some additional limited resources for analyzing single-subject data have been developed, and these are listed in Appendix D, Bibliography of Additional Resources.

What This Book Is and What This Book Isn't

This book is designed to be a helpful tool in the use of *SSD for R* for visually and statistically analyzing single-subject data. *SSD for R* has the most comprehensive functionality specifically designed for the analysis of single-subject research data currently available (Auerbach & Zeitlin Schudrich, 2013). The aim of this text is to introduce readers to the various functions available in *SSD for R* and to step them through the analytical process based on the characteristics of the collected data.

SSD for R has numerous graphing and charting functions available to conduct robust visual analysis. For example, besides the ability to create simple line graphs, additional features are available to add mean, median, and standard deviation lines to help better visualize change over time. Graphs can also be annotated with text.

Additionally, graphing functions are available to depict several non-overlapping effect sizes. Some of these, such as the Percentage of All Non-overlapping Data (PAND) and the Improvement Rate Difference (IRD), have been more difficult to calculate with larger datasets previously; however, introduction of software to graph these effect sizes has made it considerably easier to actually calculate these numeric values. This is particularly important when publishing single-subject research as calculable effect sizes are necessary in order to include these studies in meta-analyses (Morgan, 2008; Parker, 2006; Schlosser & Wendt, 2008; Scruggs & Mastropicri, 2013).

SSD for R also contains a host of statistical process control (SPC) charts that can be used to visually detect changes in the dependent variable between phases. The usage of the various available SPC charts is explained in detail in this book so that appropriate methods can be selected based on the characteristics of your data.

This book provides a thorough explanation of traditional statistical tests that have been applied to single-subject research and teaches readers how and when to apply these to their own research or practice evaluation situations. These include descriptive statistics as well as tests of statistical significance, such as *t* tests and the conservative dual criteria. Didactic material is provided on the interpretation of statistical and supporting visual output from *SSD for R* for each of these functions.

Following both the traditions and current trends in single-subject research, most functions in *SSD for R* contain both visual and statistical output. We believe that this is one of the major strengths of *SSD for R*, as a combined approach is likely to provide support for the most pragmatic interpretations of this primarily intervention-based type of research. For example, in your own research, you may discover that your study yielded statistically significant findings that may not be clinically significant or vice versa. We believe that a combination of both visual and statistical analysis is likely to lead practitioners and researchers to make the most appropriate decisions for their clients.

This book contains numerous examples using sample data that is available to you. We recommend that as you read through the book, you download these files and replicate the examples yourself in order to learn and master the *SSD for R* commands.

In addition, each chapter contains exercises that are cumulative throughout this book. In this way, skills that are learned in previous chapters are reinforced through repetition and application in further chapters.

This book, however, is not a book on single-subject research methodology. Therefore, this text does not provide an overview of topics such as different types of single-subject research designs, an in-depth discussion of measurement, or discussions of reliability and validity. There are many excellent texts and resources available to help you better understand the broader field of single-subject research methods, and many of these are listed in Appendix D. We suggest referring to these to gain a better understanding of single-subject research in general and to assist you in designing your own studies.

The Structure of This Book

We have designed this book to help novice users of *SSD for R* walk through the process of data collection and analysis. The first section of this book consists of two chapters and provides a general introduction to both single-subject research in general and *SSD for R* more specifically. Chapter 1 discusses data collection and walks through an example of creating a spreadsheet collecting hypothetical data. In this chapter, we also introduce three practice senarios that will be used to illustrate practice evaluations throughout the rest of the book. Chapter 2 provides an overview of *SSD for R*, including how to access it, open files, and begin working with this package.

The next section of this book consists of three chapters that will guide your data analysis. Chapter 3 covers the analysis of baseline data, including addressing issues specific to single-subject research data. Chapter 4 then covers visual comparisons of phase data, while Chapter 5 introduces readers to statistical comparisons of phase data.

The remaining section of this text contains chapters in special topics related to single-subject research data analysis. Chapter 6 covers analyzing group data. This is particularly useful if the client system under consideration is made up of more than one individual. Chapter 7 contains a discussion of aggregating multiple single-subject research studies using meta-analysis. Chapter 8 covers presenting findings by using *RMarkdown*, a tool readily and freely available to *R* users. Finally, Chapter 9 covers how organizations can build research capacity by supporting research activities and reducing barriers to building evidence.

Finally, this text contains a number of appendices to make using this text and *SSD for R* easier. Appendix A provides instruction on how to enter and edit directly in *R*. This is useful if you do not have access to spreadsheet software such as Excel or Google Sheets. Appendix B provides detail about each of the *SSD for R* functions grouped by the type of analysis you would want to do. This appendix can be used if you are unsure of which function to select. Appendix B contains a series of decision trees. These flowcharts will help you select functions for specific analyses based on the characteristics of your data. Finally, Appendix C provides resources that can be used to enhance your understanding of single-subject research. These include books, websites, and other resources that you might find helpful in addition to this text.

Conclusion

Single-subject research has and continues to have an important position in the areas of both research and practice evaluation. With the current trend toward evidence-based practice, we believe that the demand for and reliance on this type of research is only going to increase in the future. *SSD for R* is a software package available to help researchers and practitioners analyze single-subject research data.

This text is designed to help readers learn to use both the visual and statistical functions in *SSD for R*. Examples, screenshots, and instructions for when and how to use each of these functions help to make this book a useful tool in the interpretation of single-subject data analysis in general and in the use of *SSD for R* in particular.

1

Getting Your Data Into *SSD for R*

Introduction

In this chapter you will learn how to measure target behaviors and use Excel or other software to record and edit client data. You will then be able to import these data into *R* and use the *SSD for R* functions to analyze them. The first part of this chapter focuses on the types of data you will want to record and some common issues related to collecting these. While an overview of this material is covered in this chapter, additional resources that include these topics in depth are listed in Appendix D. The second part of this chapter shows you how to use Excel or another spreadsheet program to quickly and effectively record the data.

What to Measure

Single-subject research designs rely on repeated measures of target behaviors that are related to client problems. You can measure one or more target behaviors using Excel or any software package whose files can be saved as comma-separated values. This data can then be analyzed with *SSD for R*. There are several types of measurements you can use to accomplish this: direct behavioral observations, standardized scales, individualized rating scales, and logs.

Direct Behavioral Observations

Direct behavioral observations refer to target behaviors that are easily observed by the client or someone else. For example, Mary is an elderly client who is experiencing social isolation. Mary or someone else, such as a family member or caregiver, could easily report how many times she participates in social activities at the local senior center each week.

When measuring behaviors directly, whoever is actually collecting the information is typically counting the number of times actual events occur. Depending on the type of behavior you are measuring and the importance of time in assessing the behavior, you may want to eventually record these data as proportions. For example, let's look at the example of our socially isolated client. Data collection lasted 4 weeks while Mary was being assessed, during which time she was asked to report how many times a week she went to the senior center. The results are shown in the table below.

SSD for R. Charles Auerbach and Wendy Zeitlin, Oxford University Press. © Oxford University Press 2022.
DOI: 10.1093/oso/9780197582756.003.0002

Week	# of Days Mary Goes to the Senior Center Each Week
1	0
2	2
3	1
4	1

While you could record these data in Excel, you, as the social worker, realize that during Week 4, the senior center was only open 5 days instead of its customary seven due to a holiday schedule. To improve the validity of your measurement, then, you may want to record the proportion of the week that Mary goes to the senior center by dividing the number of days she attends by the possible number of days that she *could* attend. Your data for this time period then would be reported in Excel differently:

Week	Proportion of Days Mary Went to the Senior Center
1	$\frac{0}{7}=0$; record "0"
2	$\frac{2}{7}=0.29$; record "0.29"
3	$\frac{1}{7}=0.14$; record "0.14"
4	$\frac{1}{5}=0.20$; record "0.20"

Standardized Scales

Standardized scales are instruments that have typically been developed for use in either practice or research to measure a particular problem or phenomenon. Standardized scales usually have a consistent method for scoring, and scores can be compared across individuals (Bloom et al., 2009; Orme & Combs-Orme, 2011).

You may choose to measure one or more target behaviors with standardized scales; however, like other types of measures, there are advantages and disadvantages to using these. Some advantages include the ability to measure multiple aspects of a complex construct, the availability of reliability and validity information in many cases, and ease of use. Disadvantages may include the appropriateness of a given standardized scale for a particular practice situation and the length of time the scale may take to complete.

To more closely understand these advantages and disadvantages, let's look at the example of Mary a bit further. You, as the social worker, suspect that Mary's social isolation may be a result of depression. There are several ways to try to assess possible depression in this client, but one way would be to use the Beck Depression Inventory (BDI) (Beck, Ward, Mendelson, Mock, & Erbaugh, 1961). This is a 13- or 21-question survey that is considered both a reliable and a valid assessment of depression in adults and includes questions about various facets of depression, including mood, pessimism, and guilt (Thombs, Ziegelstein, Beck, & Pilote, 2008). While this, then, would seem like an obvious measure of depression to use with Mary, it is fairly lengthy, and it would not make sense to have her complete it daily or even weekly. With this in mind, it might make sense clinically to use the BDI to evaluate her depression and track Mary's progress over time, but for the purposes of a single-subject practice evaluation, you would probably want to use this measure in conjunction with others as you might want more frequent measures to aid in the analysis of your work with this client.

There are many sources for locating standardized scales, but one place to start is with the work of Orme and Combs-Orme (2011), who included scales and references for online and published standardized scales in two appendices.

Individualized Rating Scales

The individualized rating scales are created by the practitioner or researcher specifically to meet the needs of an individual client. Typically, these scales are brief and may consist of only one or two items. These can be used when no standardized scales are available that adequately assess the target behavior or you would prefer to use a brief measure of a target behavior.

The advantage of using individualized rating scales is that they can be used with almost any client since they are designed to specifically meet the needs of an individual. They are typically easy to administer and score, they are appropriate for measuring the intensity of a target behavior, and they are good for measuring beliefs, thoughts, and feelings. When well designed, individualized rating scales are usually considered reliable and valid when used repeatedly to measure a target behavior with the same individual (Bloom et al., 2009; Orme & Combs-Orme, 2011).

Care should be taken in designing self-anchoring scales for use in practice evaluation and/or single-subject research. Several excellent resources for designing these are listed in Appendix D.

To better understand how individual rating scales can be used in single-subject research, let's return to the example of Mary, our socially isolated, possibly depressed, senior citizen. The social worker has decided that she would like to use an individualized rating scale to assess Mary's mood and comfort with going out in public. She creates an individualized rating scale made up of two questions:

1. On a scale of 1 to 5 with 1 being the most uncomfortable you have ever been and 5 being the most comfortable you have been, how did you feel the last time you went to the senior center?
2. On a scale of 1 to 5 with 1 being the saddest you have ever felt and 5 being the happiest you have ever felt, what has your mood been like, on average, since we last met?

During Mary's assessment period, which has lasted 4 weeks, you collect the following information:

Week	Comfort at Senior Center	Mood
1	1	3
2	4	2
3	2	2
4	2	1

Note that Mary could very easily be asked these questions when she meets with her social worker, and the responses gathered could actually be used therapeutically during the session in which she reports the information.

Logs

Logs can be referred to as client-annotated records, critical incident recordings, or self-monitoring, but no matter how they are referred to, logs can be used to provide a qualitative dimension to quantitative data you collect using other means. In general, a log is an organized journal of events related to client problems. The log usually includes the clients' perceptions of these and/or the circumstances under which they occur. While we might ask a client to use a log, they can also be completed by professionals and significant others to get a deeper, more thorough understanding of the overall problem we are seeking to address. Logs, then, are personal and specific to the individual client and can be used to help formulate client-specific interventions.

The format of a log is up to the preferences of the practitioner and client. Some logs are very structured and can be in the form of a table that the client is asked to fill in. Some are very unstructured and can simply be made up of a notebook, pad of paper, or mobile device that the client uses for the purposes of recording events and his or her reactions to these.

When using logs with clients, you will want to decide whether you want them used at preset times or if they are to be open ended. When the client's problem occurs regularly, you may want to suggest that logs be used to record events and reactions at preset times. When this is done, logs can generate a lot of information due to their frequent use; however, if the problem is sporadic, regular use of logs can generate large amounts of useless information. In this case, you may suggest that your client use his or her log only when the problem occurs.

In order to maximize the reliability of logs, you should consider discussing the importance of accurate reporting with your clients and assure them that the use of these falls within the bounds of confidentiality set up in other aspects of your work. Creating a structured format for logs may help minimize difficulties in clients using them. You will also want to be sure that your clients understand what they are to record and how often. If you are asking your clients to use logs when an event occurs, it is advisable to have them record the event and their reactions as close in time to the event occurring as possible. In all cases, you will want to encourage your clients to be as accurate in their reporting as possible without inadvertently encouraging inaccurate reports of improvement.

Both Bloom, Fischer, and Orme (2009) and Orme and Combs-Orme (2011) provide more detailed information about the use of logs in single-subject designs, and both texts provide templates for logs that can be tailored to meet your clients' needs.

Using Spreadsheets to Record Client Data

Once you determine *what* you are going to measure, *how* you are going to measure it, and *how often* you are going to measure, you will want to collect and record these data so that it can eventually be imported and analyzed in *SSD for R*. In this section, we walk you through the steps necessary in order to do this accurately.

One of the simplest ways to bring your client data into *SSD for R* for analysis is by entering it into Excel, Google Sheets, Numbers, or any other program that can create ".csv" files. Because Excel is the most commonly used spreadsheet program, this chapter shows you how to enter data in Excel. Other programs used for entering data will use a method similar to, although not exactly the same as, Excel.

In some cases, you may not be able to use Excel or another program to enter your data. It is possible to enter your data directly into *R* for analysis into *SSD for R*. This is explained in detail in Appendix A, "Entering and Editing Data Directly in *R*."

If you are using Excel or another program, you will need to create a separate file for each client/group; however, you can track multiple behaviors for each client/group in a single file.

Creating a file that can successfully be imported into *SSD for R* has to be done in a particular manner. To do this, complete the following steps:

1. Create a folder that will be used to store your data and give it a name you will remember, such as *ssddata*.
2. Open Excel.
3. On the first row (labeled "1"), enter names for each behavior (i.e., variables) across the columns, beginning with column A.

   HELPFUL HINT: For each behavior you are measuring, you will need to create both a behavior variable and a corresponding phase variable. The behavior variable will measure the target behavior, and the phase variable will indicate in which phase, baseline or intervention, the measurement occurred. In order to do this systematically, we recommend giving the behavior a meaningful name, and the associated phase variable name should be similar to the behavior variable name. For example, if you are measuring crying behavior, your behavior variable could be named "cry" and your phase variable could be named "pcry."

4. Starting in row 2, begin entering your data for each occurrence the behavior is measured.

   IMPORTANT NOTE: When phases change, you will need to enter "NA" (do not enter "N/A" or "na") into the row between a change in phases for both the behavior and phase variable.

To look at a relevant example, let's return to the example of Mary, discussed previously in this chapter. You will note that the social worker ultimately collected three measurements during her assessment: Mary's attendance at the senior center, her comfort with going to the senior center, and her mood. In order to record these data in Excel, we will need to create three behavior variables and three corresponding phase variables. We will call the behavior variables "attendance," "comfort," and "mood." We will call the corresponding phase variables "pattendance," "pcomfort," and "pmood." In addition, we will also create a variable called "notes" to record any critical incidents that might have occurred.

In Figure 1.1, note that we have entered baseline, or preintervention, data for Mary. If we had begun our intervention, we would continue entering that data on Excel row 7; however, the phase variable would change as "A" generally denotes baseline data, while "B" indicates data collected during the intervention.

5. Once your data are entered into Excel, you will need to save it as a ".csv (Comma delimited)" or ".csv (Comma Separated Values)" file in your *ssddata* directory. To do this, click SAVE AS and choose a name for your file. Do NOT click SAVE, but instead select one of the .csv options from the drop-down menu for SAVE AS TYPE or FORMAT, as shown in Figure 1.2 (see p. 16). After you finish this,

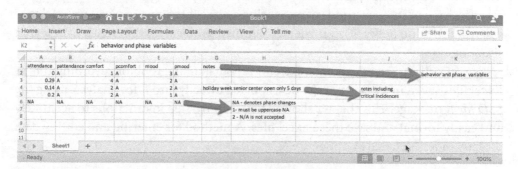

Figure 1.1 Baseline data entered into Microsoft Excel

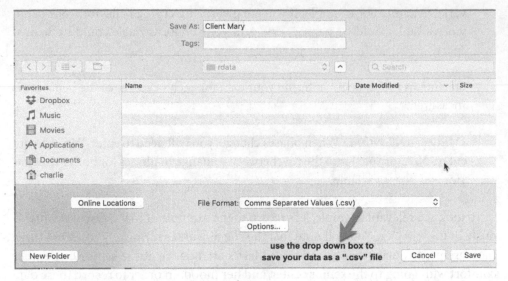

Figure 1.2 Saving a Microsoft Excel file in ".csv" format

you should click SAVE and close Excel. You may receive several warnings, but you can accept all of these by selecting CONTINUE.

Once you enter data into Excel, you can import it into *SSD for R* and begin your analysis. In the next chapter, you will learn how to access *R*, the *SSD for R* software, and *RStudio* user interface to complete this analysis.

Conclusion

In this chapter, you learned about different ways to measure target behaviors associated with your clients' identified problems. Four different types of measurement exist, each of which has its own strengths and weaknesses. These include direct behavioral observations, standardized scales, individual rating scales, and logs. When selecting one or more methods of measuring a target behavior, you will want to consider the specific needs of your client, identified problem, and practice or research situation.

As you collect your client data, Excel is an outstanding tool for recording this as Excel files saved in .csv format can be imported directly into *R* for analysis using the *SSD for R* package. This chapter showed you step-by-step instructions to accurately and easily format and save these Excel files. A method for entering and editing data directly in *R* is detailed in Appendix A.

Chapter 1

Assignment 1.1—Entering Data

Brenda is a developmentally delayed client with oppositional behavior. She was in danger of not being accepted into a residential program because she didn't get along with others. Let's examine one aspect of her behavior, oppositional behavior in school.

This was important to measure because each episode typically lasted 20–30 minutes and was disruptive to the school day.

For this assignment, you will create a Excel or Google Sheet with Brenda's baseline data. BE SURE TO STORE THIS SPREADSHEET IN A PLACE YOU CAN ACCESS LATER, as you will use it for future homework assignments. The data you need for this is in the following table:

Day	# of Oppositional Episodes
1	2
2	1
3	1
4	1
5	3
6	2
7	1

1. Create the spreadsheet and name it "Brenda—FirstLast" with FirstLast being YOUR first and last name. For example, the file would be titled "Brenda—WendyZeitlin."

2
Overview of *SSD for R* Functions

Introduction

SSD for R is a software package composed of a set of statistical functions that are designed for the analysis of single-subject research data using *R*, a free, open-source statistical programming language (R Core Team, 2013). *R*, in general, is becoming popular. There are numerous readily available, free resources to help people get started with basic and advanced use and programming; interest in *R* has virtually exploded in recent years. In this chapter, you are given step-by-step instructions on how to access the software necessary to use this package. You also are presented with a brief overview of the capabilities of the *SSD for R* package. How and when to use these packages are expanded on in subsequent chapters.

Getting Started

SSD for R is a set of functions that was designed for single-subject research in the behavioral and health sciences and is written in *R*, a programming language designed for statistical analysis and graphics. *R* provides an environment where statistical techniques can be implemented (The R Project for Statistical Computing, n.d.). Its capabilities can be extended through the development of functions and packages. Fox and Weisberg stated, "One of the great strengths of *R* is that it allows users and experts in particular areas of statistics to add new capabilities to the software" (2010, p. xiii). *SSD for R* is a software package that extends the capacity of *R* to analyze single-system research data (Auerbach & Zeitlin, 2021; Auerbach & Zeitlin Schudrich, 2013).

Throughout this book, we demonstrate statistical procedures using the *SSD for R* package. In order to get started using *SSD for R* yourself, you will need to download three things: *R*; *RStudio*, which we use as a user interface for *R*; and the *SSD for R* package (Auerbach & Zeitlin Schudrich, 2013; Free Software Foundation, Inc., 2012; R Core Team, 2013). *R* and *RStudio* can be downloaded either directly or through links from The Single-System Design Analysis website (https://www.ssdanalysis. com). Once *R* and *RStudio* have been downloaded, the *SSD for R* package can be installed using a simple *R* command.

Begin by downloading *R* and *RStudio* free of charge from the Single-System Design Analysis website (https://www.ssdanalysis.com or from the companion site to this text at Oxford University Press). When you click the links for each of these, you will be taken to external sites. Both *R* and *RStudio* are completely free and are considered safe and stable downloads.

SSD for R. Charles Auerbach and Wendy Zeitlin, Oxford University Press. © Oxford University Press 2022.
DOI: 10.1093/oso/9780197582756.003.0003

RStudio has developed a cloud version as an alternative to running *RStudio* on your computer. All you need to run the cloud version is a web browser and an *RStudio* Cloud account. You will also find instructions on using the *RStudio* Cloud version on both the Single-System Design Analysis website and the companion site to this text at Oxford University Press.

Throughout the rest of this book, we use sample datasets that you can download from https://www.ssdanalysis.com or from the companion site to this text at Oxford University Press. To download these, click on the "Datasets" tab from the home page and follow the instructions. These files include stability.csv, jenny.csv, jennyab.csv, ed.csv, and social_skill_group.csv. *R* scripts for each chapter are also provided on the website.

As you are getting started, you may want to create a single folder on your computer to hold your *R* data files. These downloaded files should be stored there.

Once these are installed, open *RStudio*. When you open it, your screen should look like Figure 2.1.

From within *RStudio,* type the following *R* command in the Console:

>**install.packages("SSDforR")** and press <ENTER>.

After the *SSD for R* package is installed, type the following in the Console:

>**require(SSDforR)** and press <ENTER>.

Another way to load the *SSD for R* package is to click the Packages tab in the lower right pane of *RStudio*. Click "Install Packages," and a dialogue box will appear. Select

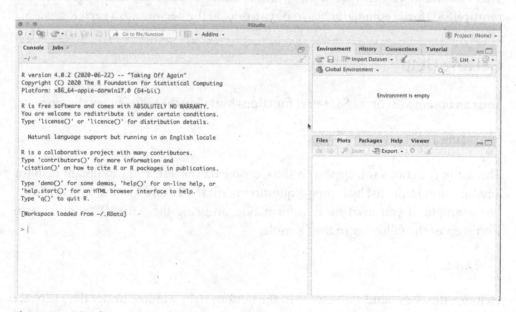

Figure 2.1 *RStudio*

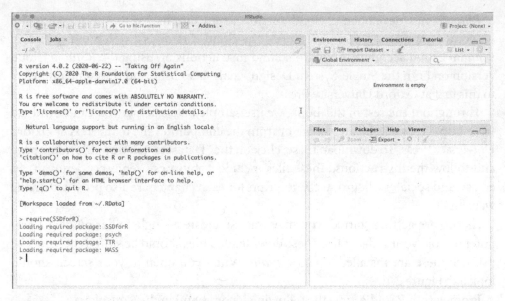

Figure 2.2 *RStudio* with SSD for R installed and required

"Repository (CRAN)" as the location to "Install from:" Under "Packages," type "SSDforR." Select this package and then click "Install." That package will automatically be installed, and the Packages tab will be refreshed. Now, check the box next to *SSDforR* to make it available for your use. You will only need to install *SSD for R* once but will need to check the box for each new *R* session.

Your screen should now look like what is displayed in Figure 2.2.

Important Note: The package only needs to be installed once using **install.packages()**. This command can also be used to install other *R* packages and upgrades. The **require(SSDforR)** command has to be run once at the beginning of each *R* session.

Getting Help

You can obtain a list of the *SSD for R* functions by typing the following in the Console:

```
>SSDforR()
```

The list of functions will appear in the Console on the left. You can obtain help for any function in the list by typing a question mark (?) followed by the function title. For example, if you need more information on using the **ABrf2()** function, you would enter the following in the Console:

```
>?ABrf2
```

Help for the command will appear in the bottom right window under the Help tab. Your screen should look as it does in Figure 2.3 (see p. 21).

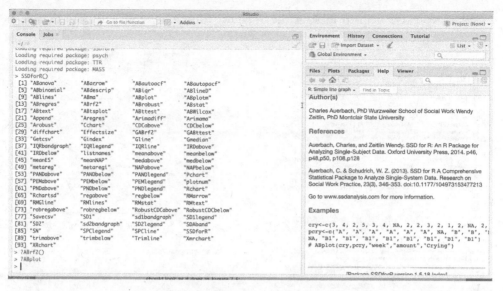

Figure 2.3 *RStudio* with functions listed and help file displayed

All help files have the same structure. They begin with a brief description of what the function does. The "Usage" section shows the required syntax for the function, and the "Arguments" section describes what is required in the syntax. The "Authors" section lists the authors of the function, and the "References" section provides additional information. At the bottom of each help file is an "Examples" section that can be copied and pasted into the Console and run. For example, type the following in the Console:

>?ABplot

When you scroll to the bottom of this help file, you will see an example. Copy the example in the help file and then paste it into the Console. Remove the number sign (#) and press <ENTER>. Your screen should look like what is displayed in Figure 2.4 (see p. 22):

You just created your first graph in *SSD for R*, and you now see in the bottom of the Console the proper syntax for the **ABplot()** command.

Alternatively, you can access hyperlinks for all *SSDforR* functions by clicking the hyperlink for SSDforR in the Packages tab in the lower right pane of *RStudio*.

A Word About the Functions

Each of the *SSDforR* functions is a statistical test and/or graph that is appropriate for single-subject data analysis. This book will teach you when and how to use each function.

When you use help to look at the *SSDforR* functions, you will note that each

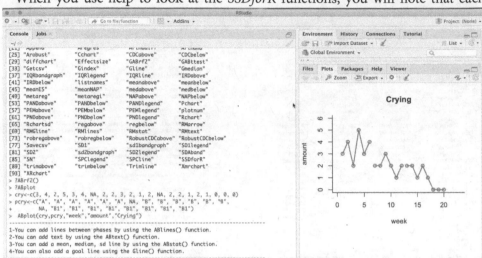

Figure 2.4　Using examples in **ABplot()** help file

function is written in a similar format. Each begins with the function name and is followed by a series of words, which we call "arguments," in parentheses. The way in which these functions are written will need to be repeated in the Console in the exact format it is shown in the Help file.

Let's begin by going over the meanings of the majority of the arguments that are specified inside the parentheses in the functions:

- *behavior*—This argument refers to the specific behavior variable that you want to analyze.
- *phasex*—This argument refers to the phase variable associated with the behavior variable you want to analyze.
- *v*—However it is written (v, v1, v2, etc.), this argument refers to the value in the phase variable that you want to include in your analysis. For example, if you want to look at baseline data, it is likely that your value for *v* would be "*A*". Every time you enter a function that includes *v*, that value should be put in quotation marks.
- *ABxlab, ABylab, ABmain*—These arguments refer to how you want your graphs labeled. *ABxlab* refers to the *x*-axis label; *ABylab* refers to the *y*-axis label; *ABmain* refers to the main title of the graph. Every time you enter a function that includes any one of these, those values should be put in quotation marks.
- *statx*—This argument refers to the statistic you would like to obtain for your data. Acceptable choices for this include "mean," "median," and "sd." Every time you enter a function that includes *statx*, that value should be put in quotation marks.

- *textx*—This argument refers to any text that you would like to insert on a graph. Every time you enter a function that includes *textx*, that value should be put in quotation marks.

Note: Most, although not all, function names begin with a capital letter. When you type the functions in the Console, you must type it EXACTLY as it appears in the help file.

Example Data: A Case Study of Jenny

There are three different example case studies referred to throughout the text: Jenny, Gloria, and Anthony. Jenny is a first-grade student at Woodside Elementary School. Her teacher, because of Jenny's disruptive behavior in class, has referred her to the school social worker. The social worker, on evaluation of Jenny, has learned that Jenny repeatedly yells at her teachers and peers, cries in class, and interrupts the class by calling out of turn frequently.

To get a better understanding of Jenny's behavior, the school social worker collects baseline data to aid in her evaluation by asking the classroom teacher to tally Jenny's troublesome behaviors. She uses Excel to record information about how many times a day Jenny yells, cries, and calls out in class. These data are shown in Figure 2.5 (see p. 24).

The column labeled "yell" shows the number of times that Jenny yelled during the school day. The label "pyell" is a phase variable associated with the variable "yell." Since there has been no intervention done at this point, the "A" for each measurement indicates that these are baseline data. The "cry" variable indicates how many times Jenny cried in school each day, and "pcry" is the phase variable associated with "cry." The "callout" indicates how many times Jenny called out in class without raising her hand, while "pcallout" is the phase variable associated with "callout." Note that each behavior variable has an associated phase variable, which is required in order to analyze these data using *SSD for R*. While we have given each phase variable a prefix of "p," you could actually name these phase variables anything meaningful to you.

To understand this, we can see that on the first day of collecting data Jenny yelled twice, called out of turn three times, and cried once. On the second day, she yelled twice, cried twice, and called out three times.

The "notes" column is optional but allows the social worker to put in his or her comments and note critical incidences that could impact any of the data. Note that the social worker mentioned that Jenny's mother was out of town one day, and, on another day, Jenny got sick while she was in school and needed to go home early. On the day Jenny went home sick, she cried 10 times, which seems to be a lot, even for her.

This Excel data file can be imported into *RStudio* so it can be analyzed using *SSD for R*. In order to do that, however, remember that the spreadsheet needs to be saved as a "CSV (Comma delimited)" file.

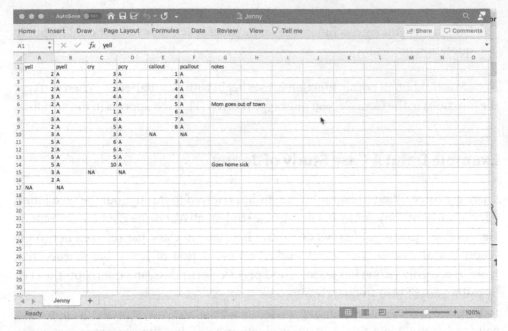

Figure 2.5 Excel spreadsheet with Jenny's baseline data

To begin your analysis, you need to load the *SSD for R* package. To do this type the following in the Console:

>**require(SSDforR)**

To import *Jenny.csv* into *RStudio*, type the following in the Console:

>**Getcsv()**

This command is one of the functions in *SSD for R*. Once you hit <ENTER>, you will be prompted to select a file. Select *Jenny.csv* and, after opening the file, type the following in the Console:

>**attach(ssd)**

As displayed in Figure 2.6 (see p. 25), to view your file in the top left pane in *RStudio*, follow these steps: Click **Environment** in the top right pane and then in the top right pane click **ssd** in the **Global Environment** window. Jenny's data will appear in the top left pane.
To obtain a list of variables, type the following in the Console:

>**listnames()**

Your *RStudio* window should look like what is displayed in Figure 2.7 (see p. 25).

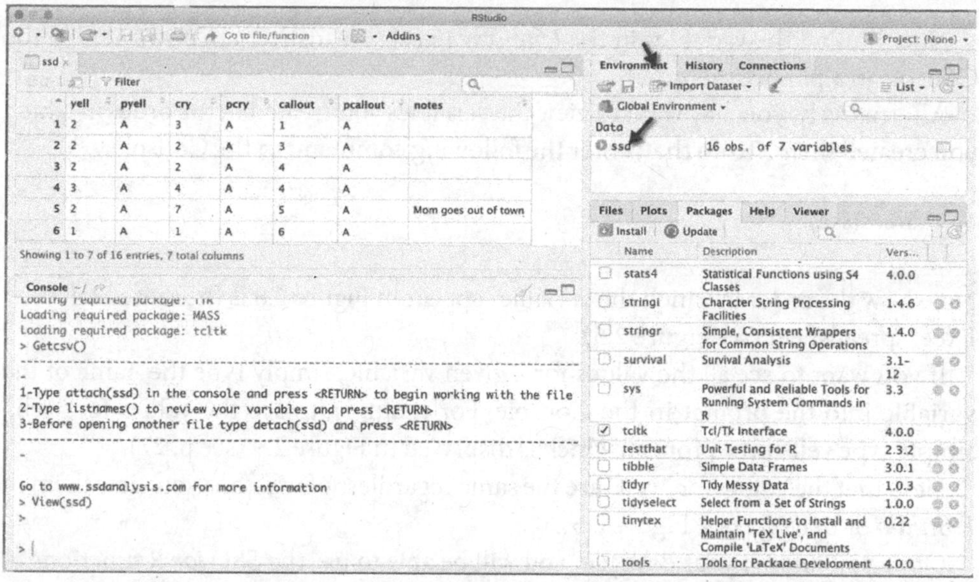

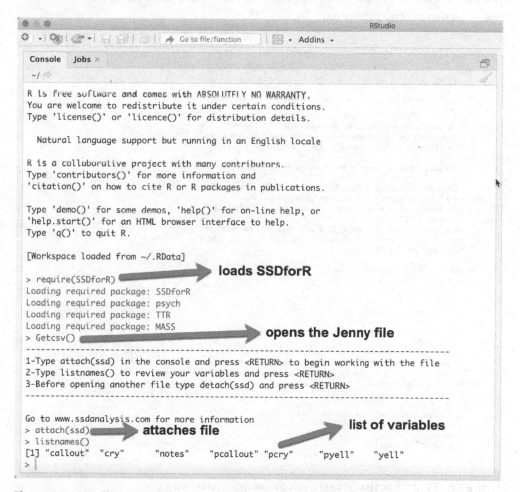

Figure 2.6 Displaying ssd spreadsheet in *Rstudio*

Figure 2.7 Using **listnames()** to view variable names in alphabetical order

Note how all the variables in the *Jenny.csv* Excel spreadsheet are now listed in the Console in alphabetical order.

Alternatively, you may want to view the variables you created in the order in which you created them. To do that, enter the following command in the Console:

>**names(ssd)**

Note how the order in which the variables appear in Figure 2.8 is the same as in your Excel spreadsheet (see Figure 2.5).

If you want to see all the values for a given variable, simply type the name of the variable into the prompt in the Console. For example, to see the values for "yell," simply type **yell** at the prompt, which is displayed in Figure 2.9 (see p. 27).

Note that the values for "yell" are the same regardless of whether you view them in *R Studio* or in Excel (see Figure 2.5).

Once you import your data file, you will be able to use the *SSD for R* functions to analyze your data visually and statistically. As you continue going through this book, you will learn more about Jenny's presenting problems by analyzing data at different points in her treatment process.

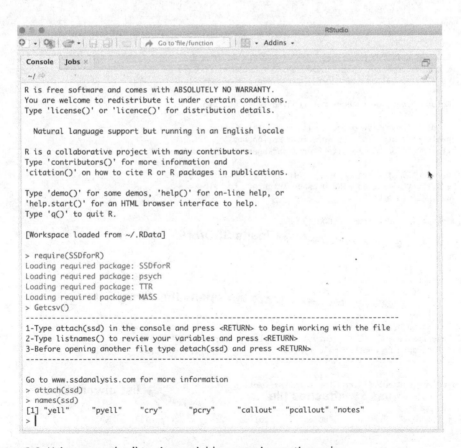

Figure 2.8 Using **names(ssd)** to view variable names in creation order

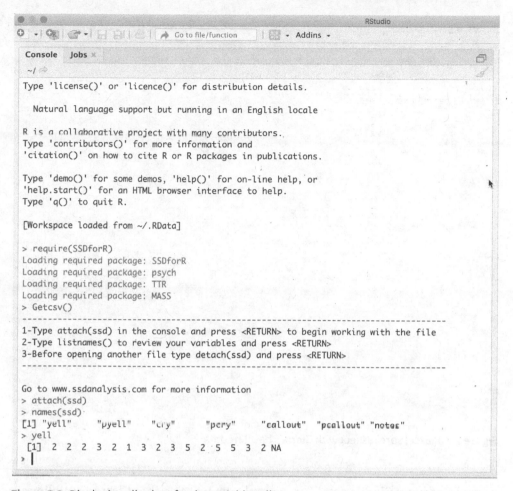

Figure 2.9 Displaying all values for the variable *yell*

Gloria a 62-year-old widow described as having a major acute depressive disorder that had its present onset after her husband's sudden death from a heart attack. Her symptoms are feeling sadness, tearfulness, lack of interest in most daily activities, insomnia, and suicidal ideations.

To better understand Gloria's degree of depression, her worker administered a depression scale daily. Also measured was the number of hours Gloria slept daily. These data are shown in Figure 2.10 (see p. 28). The column labeled "depress" shows Gloria's daily depression score. The "pdepress" is a phase variable associated with the variable "depress." Since there has been no intervention done at this point, the "A" for each measurement indicates that this is baseline data. The "sleep" variable indicates how many hours Gloria slept each day, and "psleep" is the phase variable associated with "sleep."

Anthony, age 38, was referred to therapy by his employee assistance program. Anthony reported that he was compulsively repeating some acts in his daily life. He is occupied by thoughts of repeating actions that lead to the repetition

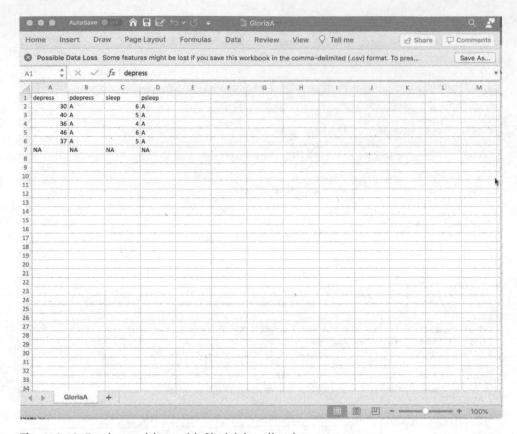

Figure 2.10 Excel spreadsheet with Gloria's baseline data

of some of his daily life acts. For example, he worries about leaving appliances on or doors unlocked; he expresses discomfort about things being out of order. To better understand Anthony's behavior, his worker recommended he keep a log of the number of times daily he checked his appliances and rearranged items. These data are presented in Figure 2.11 (see p. 29). The column labeled "checking" indicates the number of times Anthony exhibited checking behavior daily. "pchecking" is a phase variable associated with the variable "checking." Since there has been no intervention done at this point, the "A" for each measurement indicates that these are baseline data. The "arrange" variable indicates how many times Anthony rearranged items each day, and "parrange" is the phase variable associated with "sleep."

Summary of Functions

SSD for R supports the full range of analysis that can be conducted with single-subject research data. How and under what conditions are each of these functions used is discussed in detail in subsequent chapters, but a brief overview of these functions is

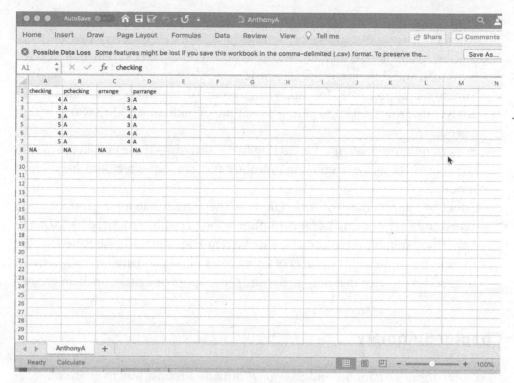

Figure 2.11 Excel spreadsheet with Anthony's baseline data

described here. You can also refer to Appendix B, *SSD for R* Functions Quick Guide, to see all of these functions grouped by usage.

- Basic graphing functions—These include a set of commands that will allow you to make and label a variety of graphs, draw vertical lines between phases, and note critical incidences. Graphs include, among other things, simple line graphs (with or without mean or median lines), standard deviation graphs, and graphs showing interquartile ranges. These graphs can be exported in a number of different formats.
- Functions associated with descriptive and basic statistics—These include basic descriptive statistics, including mean, median, standard deviations, and quartiles.
- Functions associated with statistical process control (SPC) charts—These include X-R charts, R charts, p charts, X-mr charts, and c charts. Again, all charts and graphs can be exported in a variety of formats.
- Functions associated with autocorrelation and data transformations—These are appropriate for calculating and dealing with issues of autocorrelation, which are unique to time series research designs.
- Functions associated with effect sizes—These enable users to quantify the degree of change observed between phases.

- Functions associated with hypothesis testing—These enable users to test whether there are statistically significant differences between two or more phases. Available tests include *t*-tests, chi-square, analysis of variance (ANOVA), binomial, and conservative dual criteria.
- Functions appropriate for analyzing group data—These are suitable for analyzing group data, including variation of individuals' behaviors within a group.
- Functions appropriate for conducting meta-analysis—These enable the user to conduct meta-analysis utilizing omnibus and random-effect and fixed-effect models.
- Functions appropriate for *RMarkdown*—These are suitable for producing MS Word and PDF documents for communication of results produced by *SSDforR*.
- Functions suitable for community or system-wide data—These functions are used to examine large amounts of data across phases and are typically used in time series research that may go beyond the scope of one client or client group.

Chapter Summary

In combination, *R, RStudio*, and *SSD for R* provide a robust way to analyze single-system research data. This chapter showed you how to download the necessary software and provided an overview of the visual and statistical capability available with *SSD for R*. In the following chapters, you will learn how to use these to analyze and interpret single-subject research data.

Chapter Exercises

Assignment 2.1—Opening and Displaying an *SSD for R* File

1. Load *SSDforR* and use the **Getcsv**() function to open the Jenny baseline data, attach them, and create a line graph of the yelling behavior.
2. Display the spreadsheet in the top left console of *RStudio* by double clicking *ssd* in the global environment in the top right panel. Take a screenshot of *RStudio* with the spreadsheet open.
3. You will need to submit a document (Word or Google Doc) with a copy of the graph and the screenshot.

3
Analyzing Baseline Phase Data

Introduction

In this chapter you will learn about methodological issues in analyzing your baseline. There may be times, however, when you may want to do *additional* analysis and compare, for example, one intervention phase to another. In these cases, you would also want to consider conducting the analyses we introduce in this chapter to those phases as well.

This chapter is not intended to provide a comprehensive discussion of how to design a baseline; there are a number of excellent texts listed in Appendix D that provide in-depth discussion of single-subject research designs. Rather, in this chapter we focus on assessing baseline data. The baseline is the phase in single-subject research in which data are collected on the target behavior prior to any intervention. As a result, the baseline serves as the basis for comparison with information collected during any intervention that might take place. This comparison allows you to determine the target behavior is changing and how it is changing. Well done, the baseline lets you know what would be expected to continue if there were no intervention at all (Bloom et al., 2009).

There are two different types of baselines, concurrent and reconstructed. In a concurrent baseline, data are collected while other assessment activities are being conducted. A reconstructed baseline is an attempt to approximate the naturally occurring behavior based on the memories or documentation of the client or others; case records can also be utilized to produce a reconstructed baseline. The reconstructed baseline can be utilized when a concurrent one is not possible. In some cases, a combination of reconstructed and concurrent data is used to form the baseline.

Analyzing Your Baseline

In general, analyzing baseline data is simply a matter of describing it. Traditionally, single-subject research uses visual analysis to accomplish this, and descriptive statistics, such as the mean, standard deviation, median, and quantiles, are used to provide more in-depth information. *SSD for R* can be used to accomplish this.

The other issue that is relevant in the analysis of baseline data is that of autocorrelation, also known as serial dependency. Data in any phase that have high levels of autocorrelation may need to be treated differently from data that do not have this problem when comparing phases as most analytical methods are based on the

SSD for R. Charles Auerbach and Wendy Zeitlin, Oxford University Press. © Oxford University Press 2022.
DOI: 10.1093/oso/9780197582756.003.0004

notion that observations are independent of one another. Therefore, it is important to assess the degree to which phase data are autocorrelated in order to better understand how comparisons to other phases can be made at other times, and this issue is discussed more at the end of the chapter.

Visual Analysis

Determining the stability of a baseline is an important first step in the analysis of single-system data. A stable baseline is one in which there are no wide fluctuations or obvious cycles. Stable baselines provide the ability to estimate what might happen to the behavior in the future if no intervention were provided. A lot of variability in the data makes it difficult to determine how the behavior is affecting the client system and can lead to incorrect conclusions about the impact of the intervention (Matyas & Greenwood, 1990).

Ideally, the baseline should continue until stability is achieved; however, there are situations in which this is not possible or practical (Bloom et al., 2009; Logan, Hickman, Harris, & Heriza, 2008; Portney & Watkins, 2008). In some cases, it may be necessary to intervene with the client before stability in the baseline occurs because of pressing needs of the client, and it may be possible to get a good understanding of the behavior by combining concurrent and retrospective baseline data. This issue is often the case when single-subject research is part of practice evaluation. If it is not possible to continue collecting baseline data until that phase is stable, it might be feasible to select another target behavior to assess.

Illustrations of baselines with varying degrees of stability follow next. Figure 3.1 is an illustration of a stable baseline.

Note that there is only a slight upward or downward direction in the data as all baseline values range between 3 and 4. This is a desirable pattern because there is a consistent occurrence of the behavior over time. Changes in the behavior during the intervention phase would be readily apparent.

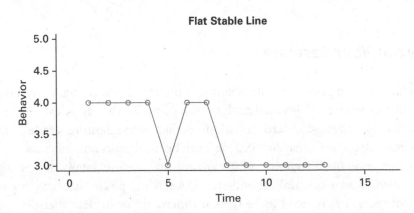

Figure 3.1 Example of a stable baseline

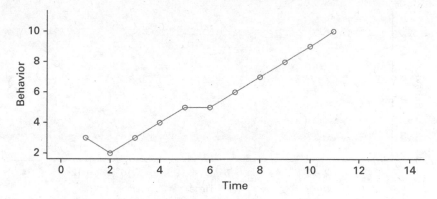

Figure 3.2 Example of an increasing stable baseline

Figure 3.2 provides an example of an increasing and stable baseline.

In this example, the baseline is stable because we can use the obvious trend to predict future values; however, the outcome being measured is consistently increasing. If the goal were to decrease or stabilize the behavior, a reversing or flattening trend in the intervention phase would indicate that. If, however, the goal was for the behavior to continue increasing, a successful intervention may appear to be a change in the rate of the increase.

Figure 3.3 shows an example of a semistable baseline.

Although a future trend can be predicted, the wide differences between data points make it difficult to draw any conclusions with certainty. Visually comparing this semi-stable baseline to future intervention data is difficult.

Finally, Figure 3.4 (see p. 34) illustrates an unstable baseline.

This is an extremely variable baseline with no ability to predict the future. Ideally, the best way to deal with this type of baseline is to continue to collect data until a

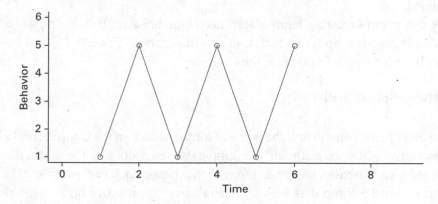

Figure 3.3 Example of a semi-stable baseline

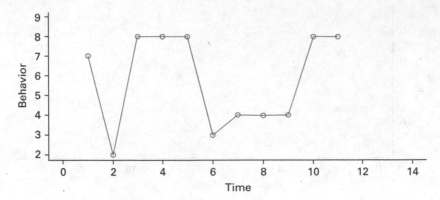

Figure 3.4 Example of an unstable baseline

pattern stabilizes. Similar to the semistable baseline of Figure 3.3, visually comparing this unstable baseline to intervention data may be difficult.

Visual Examination of Your Baseline Data: An Example

To provide a more in-depth look at analyzing baseline data, we now look at a comprehensive example that examines multiple indicators of a client problem. To begin, use **Getcsv()** to open the *jenny.csv* data file, which contains only baseline data. After the file is open, type the following in the Console:

>**attach(ssd)**

This dataset contains information about Jenny's externalizing behaviors in school, including yelling, crying, and calling out of turn as described in the previous chapter. We use these variables to assess the extent of her classroom behavior problems.

We can begin exploring Jenny's baseline yelling behavior by creating a box plot and displaying descriptive statistics, shown in Figures 3.5 (see p. 35) and 3.6 (see p. 36), by entering the following in the Console:

>**ABdescrip(yell, pyell)**

In the Plots pane, you will find the box plot for these data. In the Console, find a host of descriptive statistics, including the sample size (n), values for the mean, the 10% trimmed mean, median, sd, range, interquartile range (iqr), and quantiles. The box plot in Figure 3.5 shows data with considerable variation with a minimum value of one and a maximum of five.

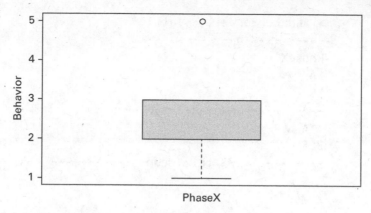

Figure 3.5 Boxplot of Jenny's yelling during baseline phase

We can also illustrate this with a simple line graph displaying the number of yelling incidents using the following command, as shown in Figure 3.7 (see p. 37):

>**ABplot(yell, pyell,"time","amount of yelling", "Jenny's yelling")**

Note: If the error "figure margins too large", occurs, you will have to increase the size of the plot pane. If this does not solve the issue, reset the plot pane by issuing the following *R* command in the Console: dev.off(). **Be mindful that any graphs currently in the pane will be removed.**

To begin our visual analysis, it is important to determine if the behavior displayed in the graph is typical (Bloom et al., 2009). One way to do this is to determine how the events vary around a measure of central tendency, such as the mean or median.

Use the following set of commands, entered into the Console one at a time, to enhance the simple line graph you created in Figure 3.7 to accomplish this, which is displayed in Figure 3.8 (see p. 38):

>**ABstat(yell,pyell,"A","mean")**
>**ABtext("mean")**

Note that when you enter the **ABstat()** and **ABtext()** functions, you will receive an additional prompt in the Console instructing you to place the mouse where you want your lines and text to appear. For the **ABstat()** command, place the cursor at the beginning left of the phase where the line should appear to ensure that the statistic line covers all the phase data. Since we only have baseline data at this point, hover your cursor in the graph above the 0 on the *x*-axis. For the **ABtext()** function, center the mouse over the area where you want the text. After the text or statistic line appears, if you are satisfied with its appearance on the graph, simply

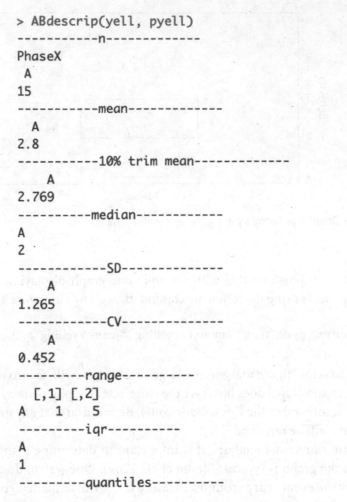

```
> ABdescrip(yell, pyell)
------------n------------
PhaseX
 A
15
-----------mean------------
  A
2.8
-----------10% trim mean------------
    A
2.769
----------median------------
A
2
------------SD-------------
  A
1.265
------------CV-------------
   A
0.452
---------range----------
   [,1] [,2]
A    1    5
---------iqr----------
A
1
---------quantiles----------
```

Figure 3.6 Console output of descriptives for Jenny's yelling during baseline phase

type "y" and <ENTER>. If you are not satisfied, type "n" and <ENTER>. In this case, you will have to reenter the command to continue annotating your graph. You can easily do this by using your up and down arrow keys to re-create previous commands.

The addition of a measure of central tendency provides valuable information. We can now see that the points in the graph vary widely from the mean and median, and the baseline looks pretty unstable. The sixth point is the furthest below the mean and median while points 10, 12, and 13 are the furthest above.

Another way to enhance your visual analysis is to examine how the data fall around standard deviation bands, which display the mean along with lines depicting the bounds of either one or two standard deviations around the mean.

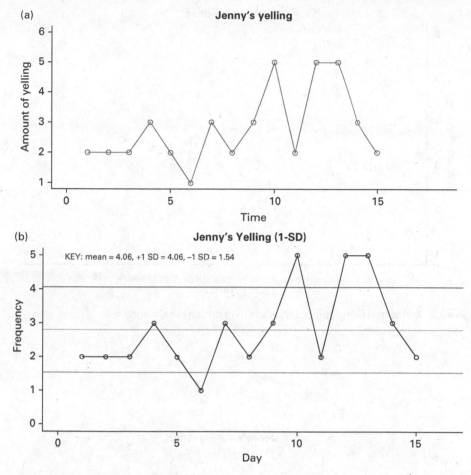

Figure 3.7 Line graph of Jenny's yelling during baseline phases

Approximately two thirds of all data in a normal distribution fall between one standard deviation (sd) above and one standard deviation below the mean. Bloom et al. (2009) suggested that, depending on the nature of the behavior, data falling outside of one standard deviation can be defined as desirable or undesirable, depending on the behavior being measured. This can be illustrated, as shown in Figure 3.9 (see p. 38), using Jenny's yelling behavior with the following command:

>sd1bandgraph(yell,pyell,"A","time","amount of yelling", "Jenny's yelling (1-SD)")

The actual values for the means and standard deviations are displayed in the Console:

"SD=" "1.26"
"+1sd=" "4.06"
"mean=" "2.8"
"-1SD=" "1.54"

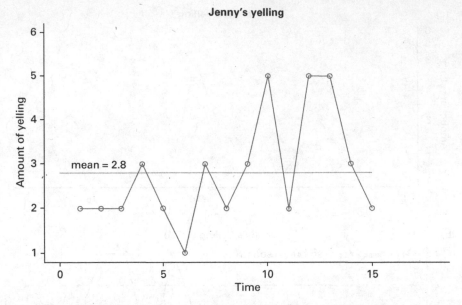

Figure 3.8 Annotated line graph of Jenny's yelling during baseline phases

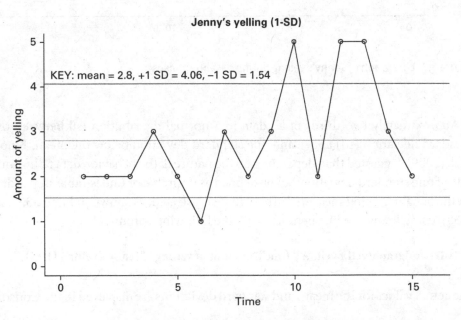

Figure 3.9 1-sd band graph of Jenny's yelling during baseline phase

and we can add this to the band graph by adding a key to your graph in an unobtrusive place:

>**ABtext("KEY: mean=2.8, +1 SD=4.06, -1 SD=1.54")**

Notice that 11 of the 15 data points in the baseline are within one standard deviation, which is 73% of the data. As this is above two thirds of the data expected in a normal distribution, we can assume that variability of this baseline is not excessively high.

Using two standard deviations would be considered even more rigorous because approximately 95% of all scores in a normal distribution are within plus or minus two standard deviations of the mean. This is illustrated in Figure 3.10, which is annotated with the values of each band and can be demonstrated using the following function:

>**sd2bandgraph(yell, pyell, "A", "time", "amount of yelling", "Jenny's yelling (2-SD)")**

As the graph displays, none of the points are above or below two standard deviations. Therefore, if we were using plus or minus two standard deviations as the criterion for typical behavior, none of the values would be considered unusual (i.e., desirable or undesirable). The actual values for the mean and standard deviations are displayed in the Console and were used in the annotation.

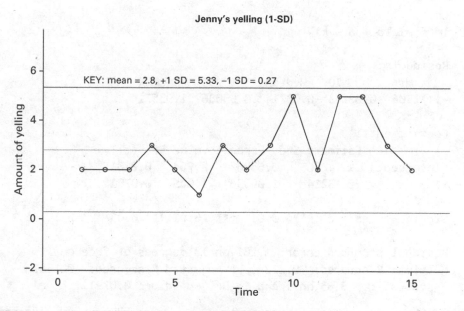

Figure 3.10 2-sd band graph of Jenny's yelling during baseline phase

"SD=" "1.26"
"+2sd=" "5.33"
"mean=" "2.8"
"-2SD=" "0.27"

Trending

As we continue our analysis of the baseline, it is important to assess if the data have a significant trend and to note the direction of the trend (i.e., whether it is increasing or decreasing). The trend can also be completely flat or irregular, as discussed previously. The goal of an intervention may be to reverse the trend in a negative behavior or increase it for positive behavior. It is important to detect a trend early on as a strong trend will impact the type of data analysis we can do in comparing phases.

As an example, let's examine Jenny's yelling behavior to see if we can detect a trend. Invoke the following command in the Console to display what is illustrated in Figures 3.11 and 3.12 (see p. 41):

```
> Aregres(yell, pyell, "A")

Call:
lm(formula = A ~ x1)

Residuals:
     Min       1Q    Median       3Q       Max
-1.72500  -0.69643  -0.00714  0.53036  1.93571

Coefficients:
            Estimate Std. Error t value Pr(>|t|)
(Intercept)  1.74286    0.63062   2.764   0.0161 *
x1           0.13214    0.06936   1.905   0.0791 .
---
Signif. codes:  0 '***' 0.001 '**' 0.01 '*' 0.05 '.' 0.1 ' ' 1

Residual standard error: 1.161 on 13 degrees of freedom
Multiple R-squared:  0.2183,    Adjusted R-squared:  0.1581
F-statistic:  3.63 on 1 and 13 DF,  p-value: 0.07911
```

Figure 3.11 Console output of assessment for trending of Jenny's yelling during baseline phase

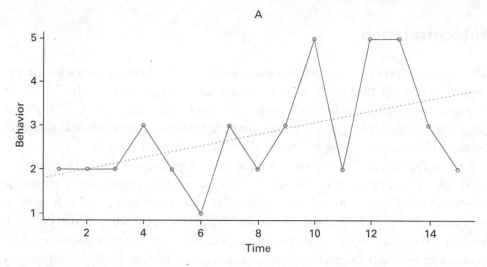

Figure 3.12 Visual assessment for trending of Jenny's yelling during baseline phase

>Aregres(yell, pyell, "A")

The graph displays a slight trend in the data, with the dotted regression line displaying the slope; however, note that the individual data points are not clustered tightly around the regression line. In addition to the graph, statistical output is produced in the Console.

These values show that the degree of change can be quantified by the estimate for x1 of 0.13214, which is the slope of the regression line. This can be interpreted as follows: For each unit increase in time, there is an estimated 0.13214 unit increase in the yelling behavior. The column labeled "t value" is the calculated value for the statistical significance of the slope and constant. The last column labeled "$Pr(>|t|)$" is the probability of making a Type I error (i.e., testing the hypothesis that the coefficient for the slope is greater than zero). Because the probability is greater than the commonly accepted 0.05 threshold ($p = .0791$), the slope is not considered statistically significant. Despite an insignificant p value, however, we may still want to consider the fit of the data around the regression line because significance may be hard to achieve, particularly with small sample sizes. If this is the case, we might want to look at the value for *Adjusted R-squared*, which explains the proportion of the variance in the outcome, in this case, Jenny's yelling, explained by the predictor variable time. In this example, we see the adjusted R-squared is 0.1581, indicating that approximately 16% of yelling is predicted by time.

A visual inspection of the graph suggests that the data are not linear and outliers at points 10, 12, and 13 have an impact on the slope by pulling it upward.

Autocorrelation

All statistical tests rely on various assumptions. When these assumptions are not met, incorrect conclusions about calculated values can be made, such as deciding that observed differences between phases are meaningful when they, in fact, are not (i.e., Type I error). On the other hand, we may erroneously not detect a difference between phases when those differences do exist (i.e., Type II error).

An assumption in the analysis of data in many types of analyses is that each observation is independent of the others. This means that observations cannot be predicted from one another. This is often the case when many research subjects are included in a study, and each individual is observed separately; however, this is less often the case when subjects may have close relationships to one another or we measure individuals repeatedly, as is the case in single-system research designs.

When data lack independence, they are considered correlated, and a special consideration in single-system research is *serial dependency* (Bloom et al., 2009). Serial dependency can be measured by testing for a correlation known as *autocorrelation* (Gast & Ledford, 2010). When visual analysis is used, high levels of autocorrelation combined with variability in the baseline increase the likelihood of a Type I error (Matyas & Greenwood, 1990).

Measuring autocorrelation using the r_{f2} (Huitema & McKean, 1994) is preferable when the number of observations are small, usually considered less than six (Bloom et al., 2009). The following command entered into the Console is needed to test for lag-1 autocorrelation for Jenny's yelling behavior:

>**ABrf2(yell,pyell,"A")**

A lag-1 autocorrelation is the degree of dependence of an observation on the previous one; a lag-2 autocorrelation is the degree of dependence of an observation on the observation that occurred two time periods prior, and so on. While there can be an autocorrelation at any lag, the lag-1 autocorrelation is considered to be the most crucial to assess (Bloom et al., 2009).

The results of the test are shown in Figures 3.13 (see p. 43) and 3.14 (see p. 43).

The r_{f2} is a measure of the degree of autocorrelation. The magnitude of autocorrelation is what is of interest here, regardless of whether the calculated value is positive or negative. The output labeled *sig of rf2* is the chances of making a Type I error testing the hypothesis that the observed autocorrelation is significantly different from zero. Because the *sig of rf2* is above 0.05 in our example with a value of 0.432, what autocorrelation we do observe is not statistically significant. As a result, we can conclude that whatever autocorrelation exists may not be problematic, and we can assume independence of observations.

There is, however, an important caveat: Because of the small number of observations in most single-system research studies, a large autocorrelation is

```
> ABrf2(yell, pyell, "A")
[1] "tf2="   "0.801"
[1] "rf2="   "0.267"
[1] "sig of rf2=" "0.432"
----------regression-----------

Call:
lm(formula = A ~ x1)

Coefficients:
(Intercept)             x1
     1.7429          0.1321
```

Figure 3.13 Console output of assessment for autocorrelation of Jenny's yelling during baseline phase

necessary to obtain statistical significance. It has been suggested that if the number of cases is small, you should assume the data are autocorrelated regardless of whether the r_{f2} is significant or not (Bloom et al., 2009). In these cases, an observed r_{f2} greater than 0.20 is considered undesirable and independence of observations should not be assumed (Bloom et al., 2009).

When assessing for autocorrelation, testing should be conducted separately for each phase. If data in any phase are autocorrelated or the number of observations is small, care should be taken in making visual, descriptive, and statistical comparisons between phases.

If you cannot assume independence of observations in any, you could attempt to transform the data in all phases in order to reduce the effect of autocorrelation. If the data have a high degree of variation, using a moving average could smooth

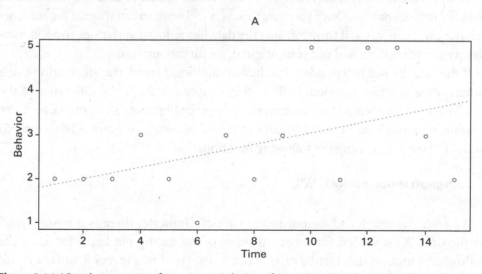

Figure 3.14 Visual assessment for auotocorrelation of Jenny's yelling during baseline phase

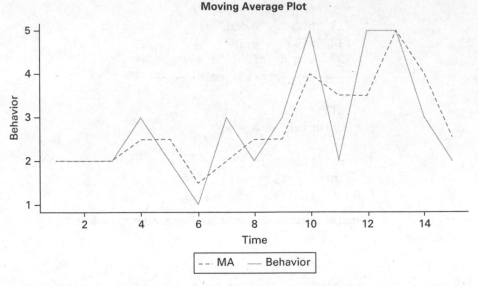

Figure 3.15 Moving average plot for Jenny's yelling during baseline phase

it. The moving average is simply the mean of two adjacent observations: ([point 1 + point 2]/2). The following function entered into the Console will perform the transformation:

>ABma(yell, pyell, "A")

You will note in Figure 3.15 that the graph does display a decrease in the degree of variation for the transformed data, which could indicate a reduction in the degree of autocorrelation.

When you use the **ABma()** function, you have an option to save the transformed data for future analysis. Once you save this, you will want to test it again for autocorrelation. If it appears as if transforming the data has reduced variation, you will want to save use these data, and not your original, for further analysis.

If the data are not independent and have a significant trend, transformation using differencing is recommended. Differencing simply calculates the difference of the value of one data point to the one immediately preceding it. As an example, begin by examining Jenny's calling out behavior at school, shown in Figures 3.16 (see p. 45) and 3.17 (see p. 45), using the following function:

>Aregres(callout, pcallout, "A")

As both the graph and output in the Console indicate, there is a strong trend in the data. Next we test for autocorrelation of the calling out behavior using the following function, the results of which are displayed in Figures 3.18 (see p. 46) and 3.19 (see p. 46):

```
> Aregres(callout, pcallout, "A")

Call:
lm(formula = A ~ x1)

Residuals:
     Min       1Q   Median       3Q      Max
-0.58333 -0.22619 -0.05952  0.19048  0.60714

Coefficients:
            Estimate Std. Error t value Pr(>|t|)
(Intercept)  0.67857    0.33651   2.017   0.0903 .
x1           0.90476    0.06664  13.577 9.91e-06 ***
---
Signif. codes:  0 '***' 0.001 '**' 0.01 '*' 0.05 '.' 0.1 ' ' 1

Residual standard error: 0.4319 on 6 degrees of freedom
Multiple R-squared:  0.9685,    Adjusted R-squared:  0.9632
F-statistic: 184.3 on 1 and 6 DF,  p-value: 9.906e-06
```

Figure 3.16 Console output for trending of Jenny's calling out during baseline phase

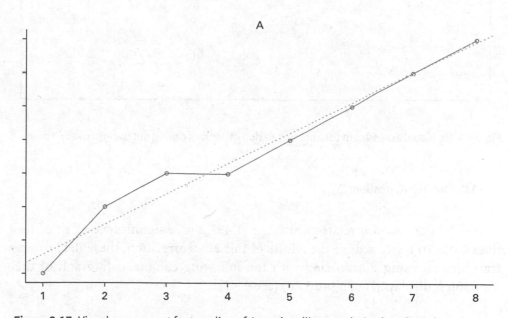

Figure 3.17 Visual assessment for trending of Jenny's calling out during baseline phase

```
> ABrf2(callout, pcallout, "A")
[1] "tf2="    "5.376"
[1] "rf2="    "1.497"
[1] "sig of rf2=" "0"
----------regression------------

Call:
lm(formula = A ~ x1)

Coefficients:
(Intercept)           x1
    0.6786       0.9048
```

Figure 3.18 Console output of assessment for autocorrelation of Jenny's calling out during baseline phase

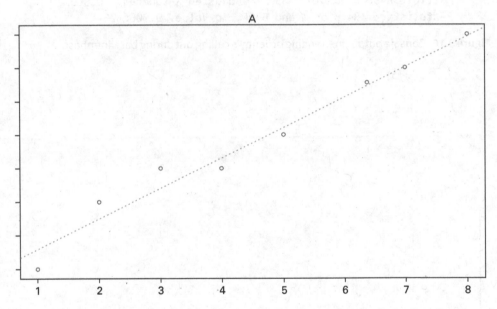

Figure 3.19 Visual assessment for autocorrelation of Jenny's calling out during baseline phase

>**ABrf2(callout, pcallout,"A")**

The data are autocorrelated with $r_{f2} = 1.497$ and a significance of r_{f2} of less than 0.05. To try to reduce the effects of this autocorrelation, these data can be transformed using differencing with the following command (a graph of this function is displayed in Figure 3.20) (see p. 47):

>**diffchart(callout, pcallout,"A")**

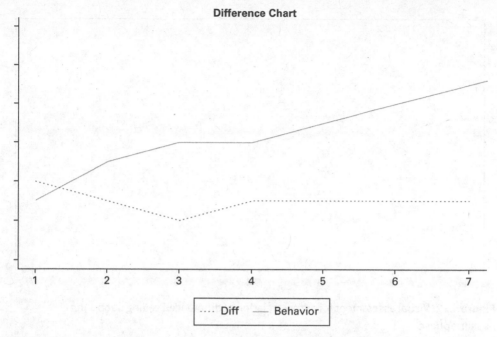

Figure 3.20 Differencing plot for Jenny's calling out during baseline phase

Since differencing seems to smooth out the line, we will save the transformed data by typing "y" when prompted.

Regardless of how you transform your data (i.e., either through the moving average or through differencing), after you have saved the transformed data, use the **Getcsv()** and **attach(ssd)** commands to open and attach the newly created dataset. Once this is accomplished, you will need to test the transformed data for autocorrelation using the following command (results are shown in Figures 3.21 and 3.22) (see p. 48):

```
> ABrf2(diff, phase, "A")
[1] "tf2=" "0.73"
[1] "rf2="    "0.458"
[1] "sig of rf2=" "0.479"
----------regression------------

Call:
lm(formula = A ~ x1)

Coefficients:
(Intercept)            x1
   1.28571      -0.07143
```

Figure 3.21 Console output of assessment for autocorrelation of differenced calling out during baseline phase

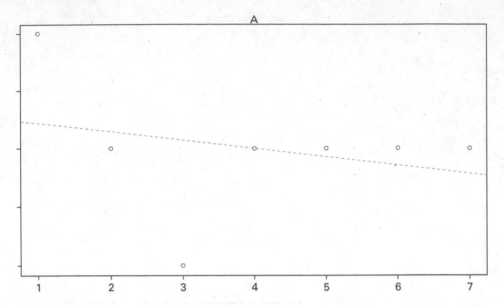

Figure 3.22 Visual assessment for autocorrelation of differenced calling out during baseline phase

>**ABrf2(diff,phase,"A")**

In the case of Jenny's calling out behavior, the transformation has reduced the r_{f2}, and it is no longer statistically significant. It should be noted that transformed data should always be tested for autocorrelation, and all other phases will need to be transformed in order to compare them.

There are situations in which transforming data does not lower levels of serial dependency. In these cases, checking for autocorrelation will yield high values for r_{f2} and/or significance values less than 0.05. When this happens, it will likely be best to continue working with the original, untransformed data as the comparison between the phases will be easier to interpret than having to account for transformed, highly autocorrelated data.

You can calculate the autocorrelation for any lag using the **ABautoacf()** function. Because of small sample size, we strongly recommend the use of the r_{f2} method to calculate the lag-1 autocorrelation. Load and attach Jenny's baseline data for an example of how to produce the autocorrelation for a lag-5 autocorrelation for her yelling behavior. Output for this is shown in Figures 3.23 (see p. 49) and 3.24 (see p. 49):

>**ABautoacf(yell, pyell, "A", 5)**

When entering this command, note that the integer "5" is the lag number requested. You can see from the correlogram in Figure 3.24 that none of the autocorrelations in lags 1 through 5 are above the upper blue line or below the lower blue line, indicating nonsignificance. The Box-Ljung test statistic is 2.9987, and the *p* value is above

```
> ABautoacf(yell, pyell, "A", 5)

Autocorrelations of series 'tsx', by lag

     0      1      2      3      4      5
 1.000  0.168  0.086  0.298 -0.123 -0.063

         Box-Ljung test

data:  tsx
X-squared = 2.9987, df = 5, p-value = 0.7002
```

Figure 3.23 Console output for assessment of lag-5 autocorrelation in Jenny's yelling during baseline

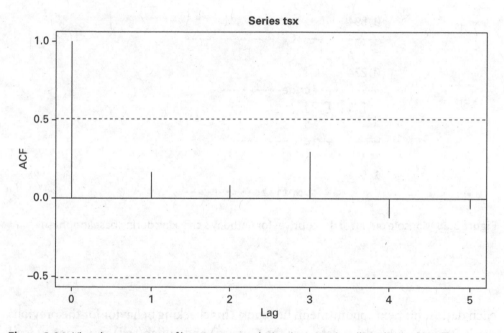

Figure 3.24 Visual assessment of lag-5 autocorrelation in Jenny's yelling during baseline

.05, further indicating that the autocorrelations for the phase are not statistically significant.

Another Example

Now, let's consider Anthony's baseline checking behavior. If you recall, Anthony is a client who appears to have obsessive-compulsive disorder, and he has been asked to count the number of times he checks that his appliances are turned off or unplugged

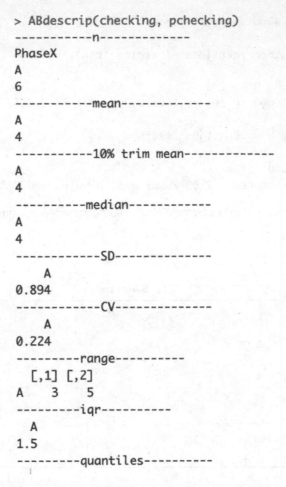

```
> ABdescrip(checking, pchecking)
-----------n-------------
PhaseX
A
6
-----------mean-------------
A
4
-----------10% trim mean-------------
A
4
----------median------------
A
4
-----------SD--------------
    A
0.894
-----------CV--------------
    A
0.224
---------range-----------
  [,1] [,2]
A   3    5
---------iqr-----------
   A
1.5
---------quantiles----------
```

Figure 3.25 Console output of descriptives for Anthony's checking during baseline phase

each day. At his next appointment, he reports his checking behavior for the previous week. To work with these data, use the **Getcsv()** and **attach(ssd)** functions to load Anthony.csv. Then, describe it as follows:

>**ABdescrip(checking, pchecking)**

The output displayed in the Console and illustrated in Figure 3.25 indicates that the mean for Anthony's checking is four times per day (sd = 0.894), with the behavior ranging from three to five times per day. The box plot, displayed in Figure 3.26 (see p. 51), shows the narrow range of this behavior, with no outliers.

With this in mind, you can create a line plot by invoking the following command:

>**ABplot(checking, pchecking, "day", "frequency", "Anthony Checking")**

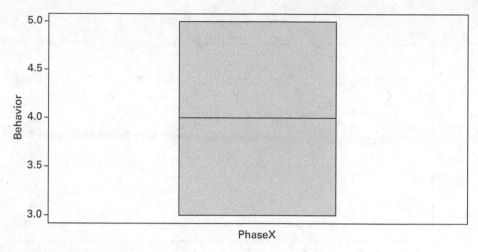

Figure 3.26 Boxplot of Anthony's checking during baseline phase

Since there were no obvious outliers in the box plot, you might choose to annotate this line graph with a labelled mean line, being sure to accept your annotations:

```
>ABstat(checking,pchecking,"A","mean")
>ABtext("mean=4")
```

This annotated graph is displayed in Figure 3.27.

From this initial analysis, it looks like the baseline is fairly stable, but we can check this out with a one standard deviation band graph, which can be annotated to display the values for the mean and the standard deviation bands:

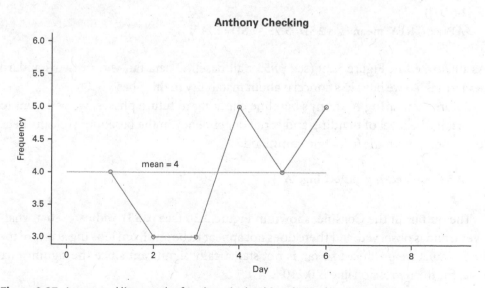

Figure 3.27 Annotated line graph of Anthony's checking during baseline phases

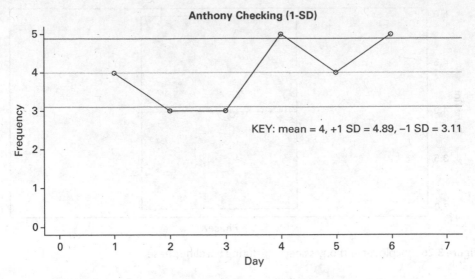

Figure 3.28 1-sd band graph of Anthony's checking during baseline phase

>**sd1bandgraph(checking, pchecking, "A", "day", "frequency", "Anthony Checking (1-SD)")**
>**ABtext("KEY: mean=4, +1 SD=4.89, -1 SD=3.11")**

This annotated graph is displayed in Figure 3.28.

Now we see that the majority of the data do not fall within one standard deviation since only two of six data points are within the bands. By expanding the graph to two standard deviations, it is hoped we will find stability:

>**sd2bandgraph(checking, pchecking, "A", "day", "frequency", "Anthony Checking (2-SD)")**
>**ABtext("KEY: mean=4, +2 SD=5.79, -1 SD=2.21")**

As illustrated in Figure 3.29 (see p. 53), all baseline data fall within two standard deviations, so we have less concern about instability in this phase.

Before comparing Anthony's baseline checking to future phases, we will want to ascertain the level of trending and serial dependency in the baseline. We can check for trending with the following command:

>**Aregres(checking, pchecking, "A")**

The output in the Console, shown in Figure 3.30 (see p. 53), indicates that whatever trend is observed, and there does not appear to be one from looking at any of the line graphs we produced so far, is not statistically significant since the significance level for the regression line is 0.2103.

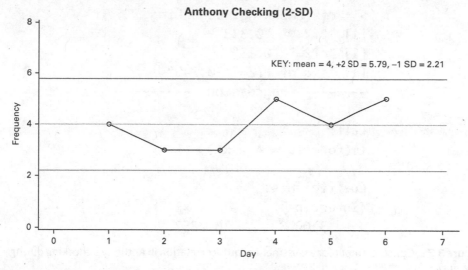

Figure 3.29 2-sd band graph of Anthony's checking during baseline phase

```
> Aregres(checking, pchecking, "A")

Call:
lm(formula = A ~ x1)

Residuals:
      1       2       3       4       5       6
 0.7143 -0.5714 -0.8571  0.8571 -0.4286  0.2857

Coefficients:
            Estimate Std. Error t value Pr(>|t|)
(Intercept)   3.0000     0.7464   4.019   0.0159 *
x1            0.2857     0.1917   1.491   0.2103
---
Signif. codes:  0 '***' 0.001 '**' 0.01 '*' 0.05 '.' 0.1 ' ' 1

Residual standard error: 0.8018 on 4 degrees of freedom
Multiple R-squared:  0.3571,    Adjusted R-squared:  0.1964
F-statistic: 2.222 on 1 and 4 DF,  p-value: 0.2103
```

Figure 3.30 Console output for assessment of trending in Anthony's checking during baseline

```
> ABrf2(checking, pchecking, "A")
[1] "tf2=" "0.333"
[1] "rf2=" "0.25"
[1] "sig of rf2=" "0.745"
----------regression------------

Call:
lm(formula = A ~ x1)

Coefficients:
(Intercept)              x1
   3.0000            0.2857
```

Figure 3.31 Console output for assessment of autocorrelation in Anthony's checking during baseline

Now we can see if there is problematic serial dependency by entering the following command:

>**ABrf2(checking, pchecking, "A")**

The output displayed in the Console and in Figure 3.31 shows a nonsignificant level of autocorrelation with a significance value of 0.745 and a relatively low level of observed autocorrelation with r_{f2}.

With no significant trend and low levels of serial dependency, we do not have to make any special considerations when comparing these baseline data to future phases at this point.

Chapter Summary

Early analysis of baseline data in single-subject research entails decision-making that informs the duration of this phase as well as how other phases may be compared to the baseline in the future. For example, visual analysis of the baseline early on may show an unstable or semistable trend that may indicate the need to collect additional baseline data in the hopes of creating stability. This analysis is critical for ensuring the integrity of the baseline for future analysis.

SSD for R enables users to gain a thorough understanding of baseline data through both visual and statistical inspection. The ability to visualize and statistically understand how data are clustered around the mean or median of the baseline (or any other phase for that matter) will give you a greater sense of the variability of the data and, in a practice evaluation setting, may guide you in setting concrete goals with clients.

In all cases, however, trending and autocorrelation of phase data need to be considered since disregard for these issues could lead to erroneous comparison of phases. This chapter provided guidance on assessment of trending and autocorrelation in small samples, which are typical of single-subject research. Additionally, *SSD for R* has the capability to transform and save phase data for which serial dependency is an issue.

Chapter Exercises

Assignment 3.1—Annotating a Line Graph

For this assignment you will continue assessing Brenda's baseline data using the file you created in Chapter 1.

1. Obtain descriptive statistics for Brenda's oppositional behaviors. In SENTENCE form, provide information about the number of observations there are in the baseline, the average number of oppositional episodes per day, and the standard deviation.
2. Create a line graph (like you did in the previous module) AND annotate it with a line denoting the mean and text, appropriately placed, to display information about the mean. Be sure to label your graph with your name in the main label.

Assignment 3.2—Baseline Analysis: Part 1

For this assignment, you will use Brenda's baseline data. Be sure to include your initials in your main labels for your band graphs. As well, be sure to answer the questions in your narrative explanation of your findings.

1. Test for baseline stability/outliers using the **sd1bandgraph()** and **sd2bandgraph()** functions. In your narrative, include the plots for each of these along with your interpretation. What percentage of data falls within each band graph? Based on this, do you think you have sufficient baseline data or would you attempt to collect more if you were able?
2. Test for a trend in your baseline data by using the **Aregres()** function. What is the slope of the trend line, and what does this mean? Is the trend significant? Would you need to consider baseline data trending in your comparison of the baseline to the intervention phase?

Assignment 3.3—Baseline Analysis: Part 2

For this assignment, you will use Brenda's baseline data ONLY.

1. Test for baseline autocorrelation using the ABrf2() functions. In your narrative, report the MAGNITUDE of autocorrelation (the rf2) and whether the observed autocorrelation is STATISTICALLY SIGNIFICANT (sig of rf2). Add to your narrative whether you think the observed autocorrelation would be problematic when eventually comparing your baseline to the intervention.

2. IF you determine autocorrelation to be problematic, what function(s) might you use to transform your data? Support your decision.

4

Comparing Baseline and Intervention Phases

Visualizing Your Findings and Descriptive Statistics

Introduction

In this chapter you will learn about methodological issues to consider in analyzing the success of your intervention and how to conduct your visual analysis. We begin with a discussion of descriptive statistics that can aid in the visual analysis of your findings by summarizing patterns of data across phases. Four common forms of descriptive statistics are explained: central tendency, variation, trends, and effect size. We then continue to a discussion of testing for autocorrelation in the intervention phase. We conclude with a discussion of the goal line, which provides a visual method to quantify effectiveness of an intervention when a specific decision is required to attain a specific goal.

Descriptive Statistics

A good way to begin comparing phases is by producing descriptive statistics for central tendency and variation for all phases. As an example, let's use the *Jennyab* dataset by invoking the **Getcsv()** function. These data are a continuation of our study of Jenny, but now includes both baseline and intervention data.

Once the dataset is open, type **attach(ssd)** in the Console to make the dataset available for analysis. Once the dataset is attached, you can produce descriptive statistics and view a box plot for both phases of Jenny's crying behavior by entering the following command in the Console:

```
>ABdescrip(cry,pcry)
```

The output, shown in Figures 4.1 (see p. 58) and 4.2 (see p. 58), includes descriptive statistics, and a box plot is displayed:

The statistics displayed in Figure 4.2 are explained as you review this chapter, but we begin with a discussion of the median. The median (Md) is the value for which 50% of the observations fall both above and below; it is the middle value if all the data were listed in increasing value. The median can be used to express the typical value in a given phase. In the Jenny example, the median number of daily episodes of

SSD for R. Charles Auerbach and Wendy Zeitlin, Oxford University Press. © Oxford University Press 2022.
DOI: 10.1093/oso/9780197582756.003.0005

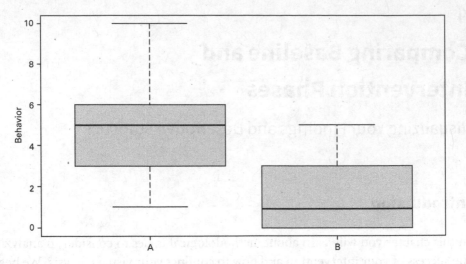

Figure 4.1 Boxplot comparing crying behavior from baseline to intervention

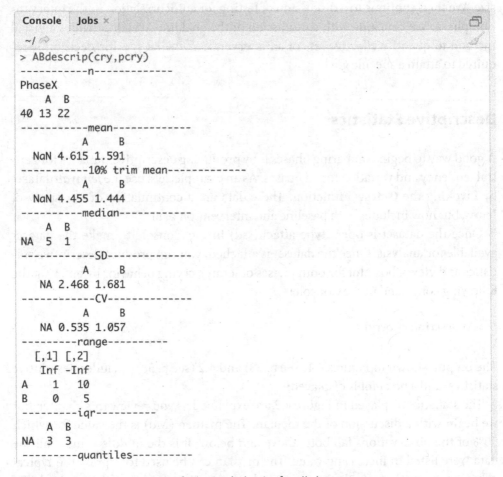

```
> ABdescrip(cry,pcry)
-----------n-------------
PhaseX
   A  B
40 13 22
-----------mean-------------
        A     B
  NaN 4.615 1.591
-----------10% trim mean-------------
        A     B
  NaN 4.455 1.444
----------median------------
  A B
NA 5 1
------------SD---------------
        A     B
  NA 2.468 1.681
------------CV---------------
        A     B
  NA 0.535 1.057
---------range----------
  [,1] [,2]
   Inf -Inf
A    1   10
B    0    5
---------iqr----------
  A B
NA 3 3
---------quantiles----------
```

Figure 4.2 Descriptive statistics for crying behavior for all phases

crying was five during the baseline (the "A" phase) compared to one during the intervention (the "B" phase). The box plot in Figure 4.1 graphically displays the median, the range, and the interquartile range for both phases. The thick black line in the boxes represents the median score for the phase. The median is visibly lower in the intervention phase, which is desirable in this case. The box plot also shows that there is considerably more variation in the baseline than in the intervention as noted by the upper and lower bounds of the box plots and the sizes of the actual boxes.

Now that we have a general sense of how the data are distributed in each phase, we can get additional information by examining a simple line graph. Consider Figure 4.3, which was produced in part by entering the following in the Console:

>ABplot(cry, pcry,"Days","Crying Episodes", "Jenny's Crying")

We can add a vertical line to separate the phases by entering the following in the Console:

>ABlines(cry)

After you type this command in the Console, click the mouse in the gap between the phases. You will be queried with a "Y/N" request to accept the line. If you are satisfied with the placement of the line, enter "Y." If you enter "N," you will then need to reenter ABlines() to place the line in the correct position.

Looking at this line graph, we note that there is an extreme value in the baseline, and that there is an associated note with that data point. In order to make this clear, we can make a notation on the graph. To do this we labeled the extreme score by entering the following command in the Console:

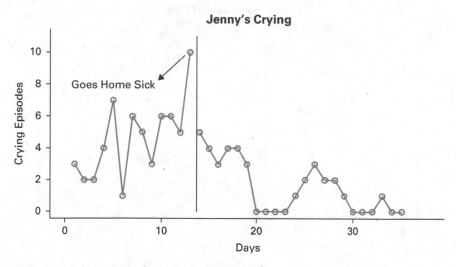

Figure 4.3 Simple line graph of Jenny's crying behavior

>ABtext("Goes Home Sick")

This is helpful in explaining why the observation is not typical. Entering the following in the Console produces an arrow extending from the outlying data point to your text:

>ABarrow()

Similar to drawing a line on the graph, you will be prompted to place the ends of the arrow in the desired locations.

Figure 4.4 illustrates that adding lines that represent the median in each of the phases can enhance the graph, especially if there are extreme, or atypical, values. To add a median line to the baseline phase, enter the following command in the Console:

>ABstat(cry, pcry, "A", "median")

Now click the mouse at zero on the *x*-axis. To add a median line to the intervention phase, enter the following in the Console:

>ABstat(cry,pcry, "B", "median")

The lines were labeled using the **ABtext()** function and clicking the mouse in the appropriate location. The addition of the median lines visually demonstrates that the typical amount of crying decreased after the intervention. The median is a good choice to use instead of the mean when there are observations that are very

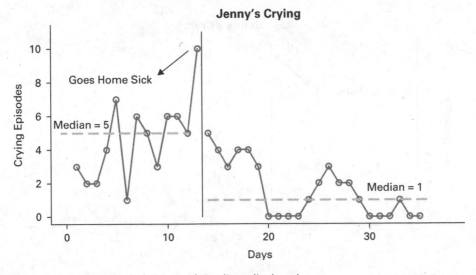

Figure 4.4 Jenny's crying behavior with medians displayed

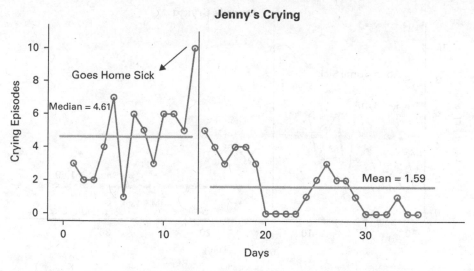

Figure 4.5 Jenny's crying behavior with means displayed

different from the typical value, or "outliers." The last observation in the baseline of 10 episodes could be considered an outlier.

The mean is another measure of central tendency that is helpful in describing the typical observation. Figure 4.5 displays a simple line graph, but this time, the means for each phase of Jenny's crying behavior are displayed. Unlike the median, the mean can be strongly impacted by outliers. This is especially true when there are a small number of observations, which is common in single-subject research.

To produce this graph, a simple line graph was first constructed using the **ABplot()** function, described previously, then the mean line was added to the baseline by entering the following command in the Console:

>ABstat(cry, pcry, "A", "mean")

A similar command, **ABstat(cry, pcry, "B", "mean")** was entered to produce the mean line in the intervention phase.

The trimmed mean (tM) can also be used when there are outliers because this measure of central tendency can dilute their impact (Bloom et al., 2009; Verzani, 2004). The trimmed mean removes a percentage of the highest and lowest values when calculating the mean. A common practice is to exclude 10% of the lowest and highest values. Figure 4.6 (see p. 62) provides an example. The trimmed mean can be added to the baseline and intervention by entering the following commands in the Console:

>Trimline(cry,pcry,"A")
>Trimline(cry,pcry,"B")

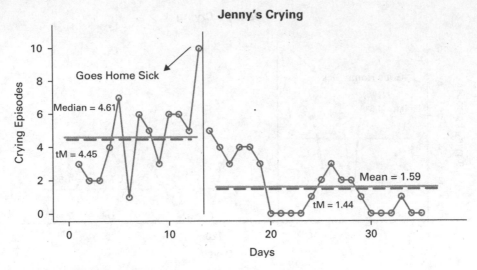

Figure 4.6 Jenny's crying behavior with means and trimmed means displayed

The text was added to baseline and intervention with the following commands:

>ABtext("tM=4.45") and ABtext("tM=1.44")

As Figure 4.6 illustrates, there is very little difference between the mean and the trimmed mean in this example. The trimmed mean is generally preferred over the median as a measure of central tendency because it uses more information about the data than the median.

Measures of Variation

Whenever a measure of central tendency (e.g., mean or median) is reported on a set of data, it is necessary to describe the degree to which scores deviate from it. Variation can be defined as the degree to which scores fluctuate around a measure of central tendency. As a result, measures of variation are necessary to fully describe a set of data. Two measures of variation that are commonly used in single-system data analysis are the range and standard deviation (SD) (Bloom et al., 2009).

Range

The range is the simplest measure of variation. It is simply the difference between the highest and lowest values in a set of observations. The range in the baseline of Jenny's crying behavior is nine, and it is five for the intervention. A more useful form of this measure is the interquartile range (IQR); the IQR is the difference between the third (75th percentile) and first (25th percentile) quartiles. These percentiles are displayed in the box plot in Figure 4.1, with the 25th percentile represented by the bottom of each box and the 75th percentile represented by the top of each box. For the Jenny data, the IQR is 3 for both the baseline (6 minus 3) and intervention (3 minus 0) phases. The IQR can be graphed using the following set of functions, which will produce Figure 4.7:

This figure is helpful in comparing the typical amount of crying in the baseline compared to the intervention. Figure 4.7 shows that typical values were between 3 and 6 in the baseline and between 0 and 3 during the intervention. This is another way to view data displayed in the box plot that is the output of the **ABdescrip()** function.

Command	Description
>ABplot(cry,pcry,"Days","Crying Episodes", "Jenny's Crying")	Produces basic AB graph
>ABlines(cry)	Draws line between phases
>IQRline(cry,pcry,"A")	Adds IQR lines to baseline
>IQRline(cry,pcry,"B")	Adds IQR lines to intervention
>ABstat(cry,pcry,"A","median")	Adds baseline median line
>ABstat(cry,pcry,"B","median")	Adds intervention median line
>ABtext("Goes Home Sick")	Adds text
>ABarrow()	Adds arrow from point 10 to text
>ABtext("Median = 5")	Adds text
>ABtext("IQR= 3 to 6")	Adds text
>ABtext("Median = 1")	Adds text
>ABtext("IQR= 0 to 3")	Adds text

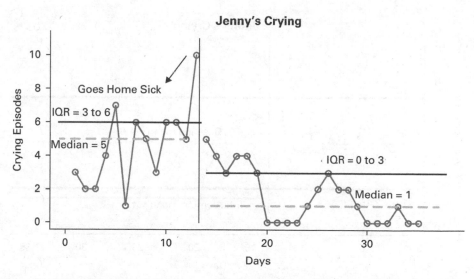

Figure 4.7 Jenny's crying with IQR ranges and medians in baseline and intervention phases

Standard Deviation

The SD is the most widely used measure of variation and is frequently used in conjunction with the mean. Like the mean, it takes into account the impact of all observations, thus using all available information. The SD is the average amount of differences of the scores from the mean. In a normal distribution, the SD provides the percentage of scores above and below the mean. In fact, if the scores are normally distributed, 68.2% of them will fall between 1 SD above and 1 SD below the mean, and 95% will fall between 2 SDs above and below the mean. "Typical" behavior can be defined as values between ±1 SD in the baseline.

Lines representing 1 or 2 SDs above and below the mean can be displayed on a graph, which provides a view of the variability between phases. Figure 4.8 displays a band graph displaying ±1 SD for the baseline going through both the baseline and intervention. The following commands will produce the graph, and output in the Console will provide values for the SD bands and the mean:

Command	Description
>SD1(cry,pcry,"A","Days","Crying Episodes","Jenny's Crying")	Produces basic graph with ±1 SD in baseline extending across phases
>ABlines(cry)	Draws line between phases
>ABstat(cry,pcry,"B","mean")	Draws mean line in intervention
>ABtext("Baseline Mean = 4.6")	Adds text
>ABtext("Intervention mean = 1.6")	Adds text
>ABtext("+1SD 7.08")	Adds text
>ABtext("-1SD 2.15")	Adds text

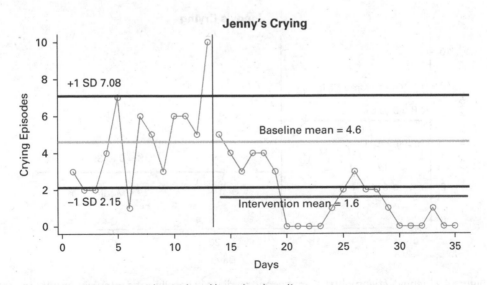

Figure 4.8 Jenny's crying with 1 SD band based on baseline

Looking at the graph in Figure 4.8, we see that a larger percentage of the scores are below 1 SD during the intervention as compared to the baseline. Since decreasing values are desirable in this example, we can assume that the intervention is making a positive difference for Jenny's crying.

How to Measure Trends

Single-subject data analysis is impacted by trends in the data. Each phase should be assessed for its presence prior to proceeding with additional inquiry (Kratochwill et al., 2013). When observations in any phase are trending significantly, either decreasing or increasing, the ability for a measure of central tendency to accurately represent the typical response is diminished as the direction in which the data are trending becomes more important. Therefore, when a trend is present, it is strongly recommended that the mean or median not be used (Bloom et al., 2009). Thus, it is a good idea to compute a trend line to help select the proper type of analysis to conduct with your data.

A trend can be defined by the slope of the best-fitting line within a phase (Kratochwill et al., 2010), also known as the ordinary least squares regression. The slope is a measure of the steepness of a line; a positive or upward slope indicates an increasing trend, while a negative or downward slope indicates a decreasing trend. With single-subject data, the slope is the average degree of change in the target behavior between time points.

Figures 4.9 and 4.10 (see p. 66) provide an example of output from entering the following command into the Console:

```
>ABregres(cry, pcry, "A", "B")
```

Note that a separate graph is produced for the two phases. The **ABregres()** function has the capability to compare any two phases of a design. If, for example, there was a B_1 phase, it could be compared to the baseline with the following command:

```
>ABregres(cry, pcry, "A", "B1")
```

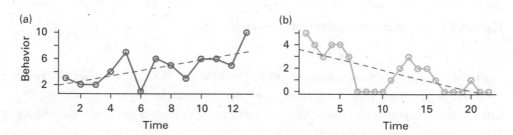

Figure 4.9 Assessing trend in baseline and intervention for crying behavior

```
 Console ~/ 
> ABregres(cry,pcry,"A","B")

Call:                                               phase
lm(formula = A ~ x1)

Residuals:                                              slope
    Min      1Q  Median      3Q     Max
-3.2033 -0.9670 -0.0275  0.8571  3.2088

Coefficients:
            Estimate Std. Error t value Pr(>|t|)
(Intercept)   1.7308     1.1520   1.502   0.1611
x1            0.4121     0.1451   2.839   0.0161 *
---
Signif. codes:  0 '***' 0.001 '**' 0.01 '*' 0.05 '.' 0.1 ' ' 1

Residual standard error: 1.958 on 11 degrees of freedom
Multiple R-squared:  0.4229,    Adjusted R-squared:  0.3705
F-statistic: 8.062 on 1 and 11 DF,  p-value: 0.0161


Call:
lm(formula = B ~ x2)

Residuals:
    Min      1Q  Median      3Q     Max
-2.3964 -0.5617  0.2515  0.9121  1.6776

Coefficients:
            Estimate Std. Error t value Pr(>|t|)
(Intercept)   3.6494     0.5490   6.647 1.8e-06 ***
x2           -0.1790     0.0418  -4.282 0.000364 ***
---
Signif. codes:  0 '***' 0.001 '**' 0.01 '*' 0.05 '.' 0.1 ' ' 1

Residual standard error: 1.244 on 20 degrees of freedom
Multiple R-squared:  0.4783,    Adjusted R-squared:  0.4522
F-statistic: 18.33 on 1 and 20 DF,  p-value: 0.0003639
```

Figure 4.10 **ABregres()** statistical output for crying behavior

Figure 4.9 displays an increasing trend in the baseline (i.e., the number of crying episodes is generally increasing from one time point to the next) while decreasing during the intervention phase (i.e., the amount of crying is generally decreasing from one time point to the next).

In the Console to the left of your graphs, you will find some important measures that quantify what is presented graphically. Here we see that the slope is positive (0.4121) compared to the intervention, which is negative (–0.1790). A decreasing trend in the number of episodes is desired during the intervention in this case as we would hope to see a reduction in the number of crying episodes. In this example, the trend is a modest one. The multiple R-squared values are 0.4229 for the baseline and 0.4783 for the intervention. The closer this value is to 1, the greater the trend and the closer to 0 is the weaker the trend. Additionally, the *p* values for the slopes of these are both statistically significant.

Now consider Figures 4.11 and 4.12 (see p. 68), which examine the trend of Jenny's calling out behavior in both the baseline and intervention phases. To generate these figures, the following command was entered in the Console:

>ABregres(callout,pcallout,"A","B")

In this example, the R squared in the baseline is 0.9685 for the baseline and 0.9388 in the intervention. In this instance, the trend line almost perfectly fits the observations. This would be considered a strong trend with slopes of 0.90476 and –0.62238 for the baseline and intervention phases respectively.

Now let's consider Figures 4.13 (see p. 68) and 4.14 (see p. 69), where we examine the trend for Jenny's yelling behavior by entering the following command in the Console:

>ABregres(yell,pyell,"A","B")

In this example, we see that the baseline has a weak trend, and the intervention has almost no trend, with R-squared values of 0.2183 in the baseline and 0.03175 in the intervention. The slopes are 0.13214 and –0.01961, respectively, which indicate fairly flat regression lines. When a trend line is flat, it indicates that the behavior is not changing over time.

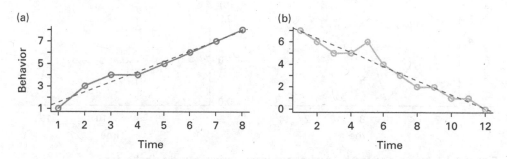

Figure 4.11 Assessing trend in baseline and intervention for calling out behavior

```
Console ~/

> ABregres(callout,pcallout,"A","B")

Call:
lm(formula = A ~ x1)

Residuals:
     Min      1Q  Median      3Q     Max
-0.58333 -0.22619 -0.05952  0.19048  0.60714

Coefficients:
            Estimate Std. Error t value Pr(>|t|)
(Intercept)  0.67857    0.33651   2.017   0.0903 .
x1           0.90476    0.06664  13.577 9.91e-06 ***
---
Signif. codes:  0 '***' 0.001 '**' 0.01 '*' 0.05 '.' 0.1 ' ' 1

Residual standard error: 0.4319 on 6 degrees of freedom
Multiple R-squared:  0.9685,    Adjusted R-squared:  0.9632
F-statistic: 184.3 on 1 and 6 DF,  p-value: 9.906e-06


Call:
lm(formula = B ~ x2)

Residuals:
     Min      1Q  Median      3Q     Max
-0.67832 -0.30594 -0.06643  0.10490  1.56643

Coefficients:
            Estimate Std. Error t value Pr(>|t|)
(Intercept)  7.54545    0.36970   20.41 1.76e-09 ***
x2          -0.62238    0.05023  -12.39 2.16e-07 ***
---
Signif. codes:  0 '***' 0.001 '**' 0.01 '*' 0.05 '.' 0.1 ' ' 1

Residual standard error: 0.6007 on 10 degrees of freedom
Multiple R-squared:  0.9388,    Adjusted R-squared:  0.9327
F-statistic: 153.5 on 1 and 10 DF,  p-value: 2.162e-07
```

Figure 4.12 ABregres() statistical output for calling out behavior

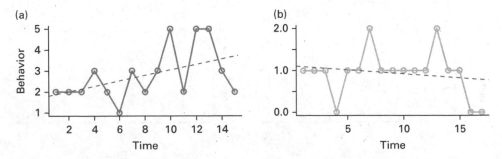

Figure 4.13 Assessing trend in baseline and intervention for yelling behavior

Using Effect Size to Describe Change

Thus far we have described visual techniques combined with measures of central tendency and variation to describe changes in behavior across phases. Effect size quantifies the *amount* of change between phases. Measuring effect size has become a common method for assessing the degree of change between phases as it gives a sense of practical, or clinical, significance (Ferguson, 2009; J. G. Orme, 1991). When effect size is used to describe data, it has advantages over tests of statistical significance that are discussed in the next section as the focus is on the magnitude of change between phases rather than whether differences are statistically significant (Kromrey & Foster-Johnson, 1996). Effect size is also relatively easy to interpret and is understandable to laypeople. There is some controversy, however, regarding how to calculate effect size (Bloom et al., 2009). *SSD for R* calculates four basic forms of effect size: ES, d-index, Hedge's g, and the g-index.

The ES (Glass, McGaw, & Smith, 1981) is simply the difference between the intervention and baseline means divided by the standard deviation of the baseline. The index can indicate no change, improvement, or deterioration. This index should be used only when there is no trend in either phase (Kromrey & Foster-Johnson, 1996). A positive ES represents improvement when an increase in the target is desirable during the intervention; a negative ES signifies improvement when a decrease in the target behavior is desirable. The ES can be interpreted in terms of the size of the difference between the phases in SD units. An ES of +1 signifies a 1-SD increase between the baseline and intervention. Conversely, a value of −1 indicates a decrease of 1-SD between the phases. The ES value can be substituted for a *z* score, which allows you to estimate the degree the behavior changed in the intervention over the baseline as a percentage (Bloom et al., 2009).

The d-index is similar to ES except that its calculation is more complex because it uses a pooled SD, which improves the accuracy of the index. Therefore, the d-index is appropriate to use when the variation between the phases differs. The interpretation of the d-index is the same as the ES; however, unlike the ES, the d-index is not directional. Like the ES, the d-index should not be used when there is trend in the data.

Hedge's g is very similar to the d-index when the number of observations is greater than 20. Hedge's g has less bias than the d-index when the number of observations is less than 20.

The *SSD for R* **Effectsize**() function automatically displays the percentage change for the ES, d-index, and Hedge's g calculations of effect size for any two phases. Bloom et al. (2009) suggested that an absolute value for ES, d-index, or Hedge's g that is less than 0.87 is a small effect size, values of 0.87–2.67 are medium effect sizes, and scores above 2.67 are large effect sizes (Bloom et al., 2009). The d-index and Hedge's g are considered more dependable than ES, which, as previously stated, is influenced by unequal variation between phases.

```
Console ~/

> ABregres(yell,pyell,"A","B")

Call:
lm(formula = A ~ x1)

Residuals:
    Min      1Q   Median      3Q     Max
-1.72500 -0.69643 -0.00714  0.53036  1.93571

Coefficients:
            Estimate Std. Error t value Pr(>|t|)
(Intercept)  1.74286    0.63062   2.764   0.0161 *
x1           0.13214    0.06936   1.905   0.0791 .
---
Signif. codes:  0 '***' 0.001 '**' 0.01 '*' 0.05 '.' 0.1 ' ' 1

Residual standard error: 1.161 on 13 degrees of freedom
Multiple R-squared:  0.2183,    Adjusted R-squared:  0.1581
F-statistic:  3.63 on 1 and 13 DF,  p-value: 0.07911


Call:
lm(formula = B ~ x2)

Residuals:
    Min      1Q   Median      3Q     Max
-1.03922 -0.07843  0.03922  0.11765  1.13725

Coefficients:
            Estimate Std. Error t value Pr(>|t|)
(Intercept)  1.11765    0.28650   3.901  0.00142 **
x2          -0.01961    0.02796  -0.701  0.49386
---
Signif. codes:  0 '***' 0.001 '**' 0.01 '*' 0.05 '.' 0.1 ' ' 1

Residual standard error: 0.5648 on 15 degrees of freedom
Multiple R-squared:  0.03175,   Adjusted R-squared:  -0.0328
F-statistic: 0.4918 on 1 and 15 DF,  p-value: 0.4939
```

Figure 4.14 **ABregres()** statistical output for yelling behavior

As an example, let's examine the effect size between baseline and intervention for Jenny's yelling behavior. We can use these measures of effect size for yelling because we saw from our previous analysis that yelling had no trend in either the baseline or the intervention. To do this, we enter the following command in the Console:

>Effectsize(yell,pyell,"A","B")

Figure 4.15 displays the results of this.

Notice the request to choose one of the following: (s)ave, (a)ppend, or (n)either results? (s/a or n), select **n**. The other two choices are used to store the effect size results to a database for use in meta-analysis. This option is discussed more fully in the meta-analysis chapter. The ES (1.46953), d-index (1.94711), and Hedge's g (1.89795) indicate a moderate degree of change. The percentages of change for the ES, d-index, and Hedge's g are similar, with values of –42.92%, 47.42%, and 47.11%, respectively. Consider that except for ES, the 95% confidence intervals range from a medium to a large effect. In other words, there is 95% confidence that the true effect size is between a medium and large effect. The finding provides some evidence that the intervention is having the desired effect.

As another example, let's examine the effect size between baseline and intervention for Anthony's obsessive checking behavior (Figure 4.16 [see p. 72]). Open the *AnthonyAB* file by typing **Getcsv()** in the Console. We can use these measures of effect size for *checking* because, as with the previous example, *checking* had no trend in

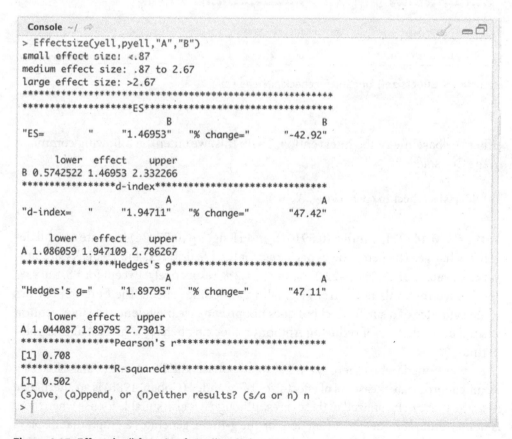

Figure 4.15 Effectsize() function for yelling behavior

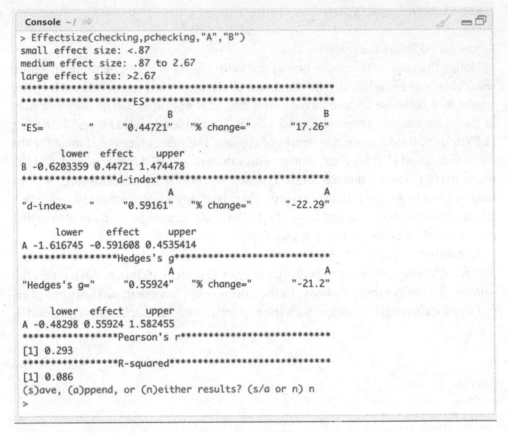

Figure 4.16 Effectsize() function for checking behavior

either the baseline or the intervention. To do this, we enter the following command in the Console:

>Effectsize(checking,pchecking,"A","B")

The ES (0.44721), d-index (0.59161), and Hedge's g (0.55924) indicate a small degree of change. The percentages of change for the ES, d-index, and Hedge's g are similar with values of 17.26%, –22.29%, and –21.2%, respectively. Except for ES, the 95% confidence intervals range from a small to a medium effect. The finding provides some evidence of a small effect but does not provide clear evidence the intervention has the desired effect of reducing Anthony's checking behavior, at least at this point in time.

As mentioned, when there is a trend in any phase, the ES, d-index, and Hedge's g are not appropriate measures of effect size. The g-index (Cohen, 1988) is an alternative measure that can be utilized in these cases, although it is also suitable for use when there is not a trend. The *SSD for R* function for the g-index generates a graph that displays the regression line, the mean, and the median for the baseline (Figure 4.17 [see p. 73]).

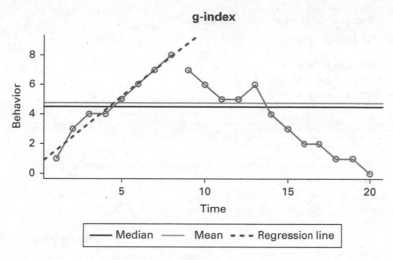

Figure 4.17 Gindex() function for calling out behavior

When there is a trend in the data, you will want to pay attention to the regression, and in the absence of a trend, it is more appropriate to focus on the mean or median. If an increase in behavior is desirable, then a higher proportion of scores above the line of focus in the intervention compared to the baseline is desirable. On the other hand, if lower scores are desirable, then a higher proportion of scores below the line would be desirable.

Consider Figures 4.11 and 4.12, where we examine the effect size for Jenny's calling out behavior. If you recall from our previous examination of trend, there is a strong trend in both phases for *calling out*. Therefore, it is appropriate to use the regression line to calculate the g-index. The index is calculated using the proportion of scores in the desired zone, in this example below the regression line, for the baseline and intervention. Then the baseline average is subtracted from the intervention average. To complete this example yourself, type the following command in the Console:

```
>Gindex(callout,pcallout,"A","B")
```

To the left in the Console, you will see the presentation in Figure 4.18 (see p. 74).

The output in Figure 4.18 displays calculations below and above for the mean, median, and regression line. In this case, the desired zone is below the regression line (G Regression line). Because the g-index is a positive number, 0.375, improvement is noted during the intervention. A positive g-index indicates improvement, while a negative value denotes deterioration.

The absolute value for the g-index can be interpreted as follows: Scores between 0.1 and 0.3 are considered a small effect, scores between 0.31 and 0.5 are considered a medium effect, and scores 0.51 and higher are considered a large effect (Cohen, 1988). Therefore, we can interpret the results to indicate that in this case there was moderate improvement between phases.

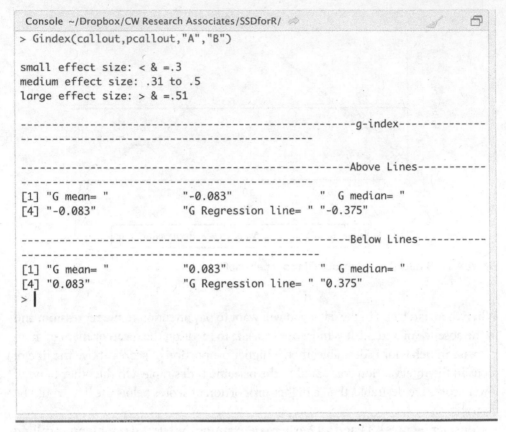

Figure 4.18 Statistical output from the **Gindex()** function for calling out behavior

Now let's look at the use of the g-index with Jenny's yelling, which you may recall only contained a weak trend in both the baseline and intervention. Figures 4.19 (see p. 75) and 4.20 (see p. 75) show results for this, which were created by entering the following command in the Console:

>Gindex(yell,pyell,"A","B")

Because there is some variation in the baseline, the median can be used to define the desired zone. Since lower scores are desirable, the desired zone is below the median. The g-index based on scores below the median is 0.816, which indicates a large degree of improvement in the behavior.

A word of caution about the use of effect sizes in general. While these calculations indicate the amount of change between phases, observed changes cannot be attributed to causality. That is, effect sizes alone, even large effects, do not prove that the intervention was the cause of the observed change.

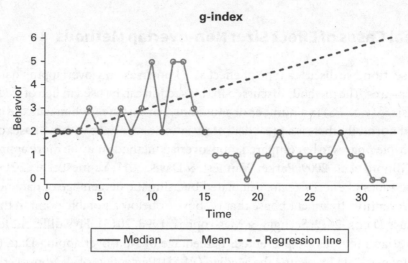

g-index

Figure 4.19 **Gindex()** function for yelling behavior

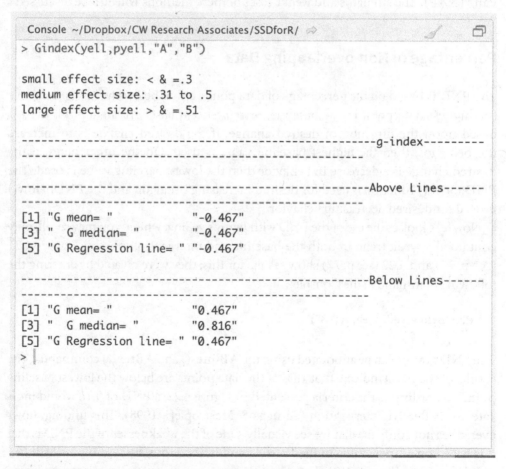

```
Console ~/Dropbox/CW Research Associates/SSDforR/

> Gindex(yell,pyell,"A","B")

small effect size: < & =.3
medium effect size: .31 to .5
large effect size: > & =.51


------------------------------------------------------g-index---------
-------------------------------------------------------------

-------------------------------------------------------Above Lines------
---------------------------------------------------------------

[1] "G mean= "              "-0.467"
[3] "  G median= "          "-0.467"
[5] "G Regression line= " "-0.467"

-------------------------------------------------------Below Lines------
-------------------------------------------------------------

[1] "G mean= "              "0.467"
[3] "  G median= "          "0.816"
[5] "G Regression line= " "0.467"
>
```

Figure 4.20 Statistical output from the **Gindex()** function for yelling behavior

Special Cases of Effect Size: Non-overlap Methods

In this section, we discuss a type of effect size known as non-overlapping data effect size measures. The methods discussed in this section can be used in lieu of traditional effects sizes (e.g., ES) in a number of situations. For instance, when data are not distributed normally, there is a great deal of variation between or within phases or there are multiple phases to be compared, non-overlap methods may be more appropriate to use (Bloom et al., 2009; Parker, Vannest, & Davis, 2011; Vannest et al., 2013).

These types of effect size are calculated by using the percentage of data points in the intervention/treatment phase that is above or below a notable point in the baseline phase (Lenz, 2012; Scruggs & Mastropieri, 1998, 2013). Five different forms of non-overlap methods are presented: Percentage of Non-overlapping Data (PND), Percentage of Data Exceeding the Median (PEM), Percentage of all Non-overlapping Data (PAND), Improvement Rate Difference (IRD), and the Non-overlap of All Pairs (NAP). The strengths and weaknesses of these methods will also be discussed.

Percentage of Non-overlapping Data

The PND is based on the percentage of data points either below the lowest or above the highest data point in the baseline, or reference, phase. The data point used is based upon the direction of desired change. If the desired change is to increase the behavior, then the highest baseline value is used. On the other hand, if the desired change is a decrease in behavior then the lowest baseline value is used. The **PNDabove()** function is used for desired increasing behavior, and the **PNDbelow()** is used for desired decreasing behavior.

Now let's look at the use of the PND with Jenny's yelling, which, you may recall, only contained a weak trend in both the baseline and intervention phases. Figures 4.21 (see p. 77) and 4.22 (see p. 77) show results for this; they were created by entering the following command in the Console:

```
>PNDbelow(yell, pyell,"A","B")
```

The PND graphs can be annotated using the **ABlines()** and **ABtext()** commands. The results of the PND indicate that 18% of the data points are below the lowest baseline point. According to the criteria presented in Figure 4.22, a PND of 0.18 would indicate an ineffective intervention (Scruggs & Mastropieri, 1998). This finding, however, does not confirm what we see visually. One of the weaknesses of the PND is that it is based on a single value in the baseline and is, therefore, susceptible to the influence of outliers (Lenz, 2012). Therefore, the presence of outliers increases the likelihood of a small effect, which may not be an accurate representation of actual change.

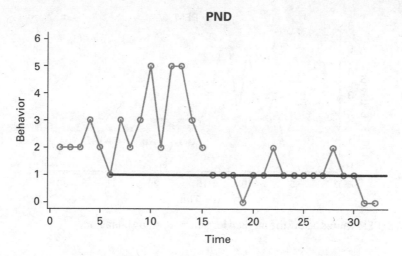

Figure 4.21 PND values below the lowest data point for Jenny's yelling behavior

```
Console ~/
> PNDbelow(yell, pyell,"A","B")
[1] "PND Below = " "0.18"
---------------------------------------------
.90 or above = very effective
.70 to .89 = moderate effectiveness
.50 to .69 = debatable effectiveness
 below .50 = not effective
>
```

Figure 4.22 Statistical output from **PNDbelow()** for Jenny's yelling behavior

Percentage of Data Exceeding the Median

The PEM procedure offers a method to adjust for the influence of outliers in the baseline phase that were observed in the PND example (Lenz, 2012; Ma, 2009). Let's look at the use of the PEM with Jenny's yelling behavior to see if using the median provides a more accurate representation of change. Figures 4.23 (see p. 78) and 4.24 (see p. 78) show results for this; they were created by entering the following command in the Console:

>**PEMbelow(yell, pyell,"A","B")**

The PEM graphs can be annotated using the **ABlines()** and **ABtext()** commands. The results of the PEM indicate that 88% of the data points are below the baseline

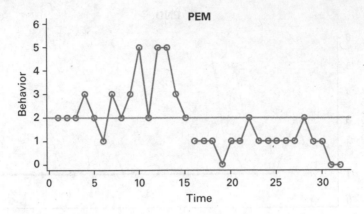

Figure 4.23 PEM values below the median for Jenny's yelling behavior

```
Console ~/

> PEMbelow(yell,pyell,"A","B")
[1] "PEM Below = " "0.88"
--------------------------------------------
.90 or above = very effective
.70 to .89 = moderate effectiveness
.50 to .69 = debatable effectiveness
 below .50 = not effective
> |
```

Figure 4.24 Statistical output from **PEMbelow()** for Jenny's yelling behavior

median of two. According to the criteria presented in Figure 4.24, a PEM of 0.88 would indicate a moderate effect size and is more consistent with what we see visually.

Percentage of All Non-overlapping Data

The PAND (Parker, Hagan-Burke, & Vannest, 2007) provides an alternative to PND and PEM, both of which stress a single data point in the baseline, but to differing degrees. The PAND uses a ratio based on non-overlap of data between phases (Lenz, 2012). Unlike PND and PEM, the PAND uses all available data from both phases (Parker et al., 2007).

Let's continue to use Jenny's yelling behavior to look at the use of the PAND. In this example, because decreasing the degree of yelling behavior is desired, the **PNDbelow()** function is utilized. Figures 4.25 (see p. 79) and 4.26 (see p. 79) show results for this; these were created by entering the following command in the Console:

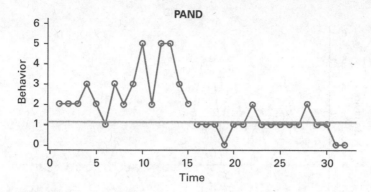

Figure 4.25 PAND values below the lowest data point for Jenny's yelling behavior

>PANDbelow(yell,pyell,"A","B")

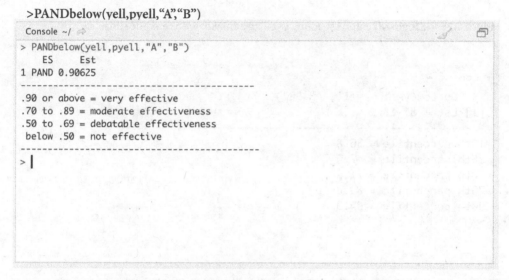

Figure 4.26 Statistical output from **PANDbelow()** for Jenny's yelling behavior

The effect size resulting from the **PANDbelow()** is 0.90625, which is considered a very effective intervention based on the criteria presented in the output.

Improvement Difference

The IRD is defined as the difference in the rate of desired behavior between the baseline and intervention phases (Parker, Vannest, & Brown, 2009).

Using Jenny's yelling behavior as an example, we can see how this can be done using the **IRDbelow()** function with yelling behavior. Here is an example:

>IRDbelow(yell, pyell,"A","B")

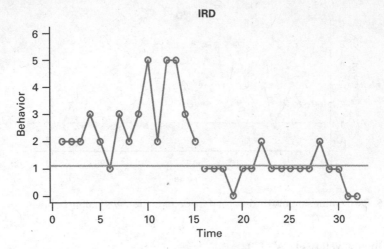

Figure 4.27 IRDbelow() graph for Jenny's yelling behavior with reference line

```
Console ~/ 🔎                                               🗗
> IRDbelow(yell, pyell,"A","B")
[1] Est = 81.18 %
--------------------------------------------
10th percentile = 36.8
25th percentile = 47.9
50th percentile = 71.8
75th percentile = 89.8
90th percentile = 99.9
> |
```

Figure 4.28 Statistical output from **IRDbelow()** for Jenny's yelling behavior

The graph in Figure 4.27 will appear.

Figure 4.28 displays the statistical output from the Console. The results indicate an 81.18% improvement from baseline to intervention. A percentile chart is presented that indicates the percentile the IRD is in (Parker et al., 2009). This chart is based on a nonrandom study of data from 166 A-B single-subject design studies (Parker et al., 2009). An IRD of 81.57% would be considered moderately high since only about 35% of IRD values obtained from the sample of studies reported by Parker et al. (2009) would be larger.

Non-overlap of All Pairs

The NAP effect size summarizes data overlap between each phase baseline data point and each intervention data point. A non-overlapping pair will have a phase

```
Console ~/
> NAPbelow(yell,pyell,"A","B")
   ·ES      Est         SE CI_lower CI_upper
1 NAP 0.9411765 0.03823692 0.7773566 0.9850379
----------------------------------------------
.93 or above = very effective
.66 to .92 = moderate effectiveness
 below .66 = not effective
----------------------------------------------

(s)ave, (a)ppend, or (n)either results? (s/a or n)
```

Figure 4.29 Statistical output from **NAPbelow()** for Jenny's yelling behavior

intervention data point in the desired direction (lower or higher) than its paired baseline phase data point. Tied pairs receive half a point. NAP equals the number of comparison pairs showing no overlap divided by the total number of comparisons (Parker et al., 2011; Pustejovsky, 2019).

Using Jenny's yelling behavior as an example, we can see how this can be done using the **NAPbelow()** function with yelling behavior. Here is an example:

>NAPbelow(yell, pyell, "A", "B")

Figure 4.29 displays the statistical output from the console.

Notice the prompt in the Console to choose one of the following: (s)ave, (a)ppend, or (n)either results? (s/a or n), select **n**. The other two choices are used to store the NAP results to a database for use in meta-analysis. This option is discussed more fully in the meta-analysis chapter.

A chart of the degree of effectiveness is presented alongside the computed effect size. In this case, the calculated value of 0.9411765 displayed below "Est" would be considered very effective (Parker & Vannest, 2009). Unlike the previous non-overlap methods discussed, dependable 95% confidence intervals (CIs) can be calculated. In this example, the CI ranges from 0.7773566 to 0.9850379, which indicates that there is 95% confidence that the impact of the intervention is at least a moderate one.

Use of Non-overlap Methods

The PND, PEM, PAND, IRD, and NAP have a number of advantages over the ES, d-index Hedge's g, and g index. A normal distribution or equality of variance between phases is not required, which makes their use more practical. Another advantage is that they can be used to compare studies containing multiple phases; for example,

A can be compared to B_1, and A can be compared to B_2 (Bloom et al., 2009; Vannest et al., 2013).

If autocorrelation is problematic, its effects can be removed by transforming the data prior to analysis. Parker (2006), however, found that the impact of autocorrelation on effect size was minimal (Parker, 2006). Manolov (2011) found that NAP is the most desirable test when autocorrelation is present. Archer (2019), on the other hand, found that the NAP was impacted by autocorrelation (Archer, Azios, Müller, & Macatangay, 2019). As a result, we strongly recommend testing for autocorrelation, and if the data are autocorrelated, to consider transforming the data (Manolov, et al., 2011). As mentioned, outliers in the data can lead to an increase in the likelihood of small effects. As a result, when there are outliers, the PEM is more appropriate than the PND.

The Goal Line

The goal line provides a visual method to quantify effectiveness of an intervention when a specific decision is required to determine if a specific objective of an outcome measured has been attained. Let's consider Gloria, who is being treated for depression at an inpatient hospital. While waiting to start a dialectical behavior therapy (DBT) program, Gloria's depression is assessed with a standardized depression scale. Conventional cutoffs for the scale are 0–9 for normal range, 10–18 for mild-to-moderate depression, 19–29 for moderate-to-severe depression, and 30–63 for severe depression. The objective is to have Gloria's score on the depression scale to at least fall within the mild-to-moderate range (below 19).

As an example, let's use the *GloriaAB* dataset by invoking the **Getcsv()** function. These data include both baseline and intervention data.

Once the dataset is open, type **attach(ssd)** in the Console to make the dataset available for analysis. Once the dataset is attached, you can produce an **ABplot()** plot for Gloria's outcomes on the depression scale by entering the following command in the Console:

```
>ABplot(depress,pdepress,"Time", "Behavior", "Goal Line")
```

and the graph will display in the Plots pane.

Next a goal line is placed on the graph by typing **Gline()** in the console. The following message will appear in the console: **Y ordinate for your goal line**. Type 19 and press <ENTER>. A goal line will be placed on the graph.

We can add a line to separate the phases by entering the following in the Console:

```
>ABlines(depress)
```

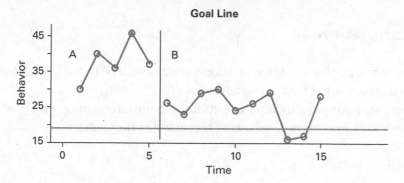

Figure 4.30 Goal line graph for Gloria's depression score

The phases were labeled using the **ABtext()** function and clicking the mouse in the appropriate location. Figure 4.30 displays the ABplot in the Plots pane.

Figure 4.31 displays a substantial decrease in Gloria's score on the depression scale. Yet, only measures on the 13th and 14th day are within the mild and moderate range of depression. Let's examine the findings more closely applying the **ABdescrip()** function by typing the following example:

>**ABdescrip(depress pdepress)**

Figure 4.31 displays a box plot in the Plots pane. The thick black line in each box represents the median for each phase (Md_a = 37 and Md_b = 26). Although an 11-point decrease in the median from the baseline to the intervention is substantial, it can lead to a different conclusion than observed in Figure 4.30, that the DBT is not efficacious.

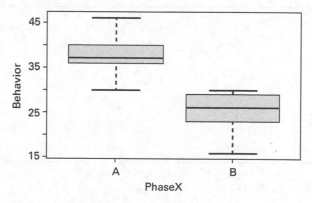

Figure 4.31 Box plot of Gloria's depression score

Autocorrelation

As stated in the previous chapter, autocorrelation should be tested in each phase as it can impact the results of tests of significance.

We can, for example, look at autocorrelation in the intervention phase for Jenny's crying behavior by typing the following command in the Console:

>ABrf2(cry, pcry, "B")

Statistical results are displayed in Figures 4.32 and 4.33 (see p. 85).

The r_{f2}, the measure of the degree of autocorrelation, is high and statistically significant, thus indicating that the observed autocorrelation may be problematic. This will have to be considered in any decisions made to test for statistical differences between phases.

We can try to transform the data to reduce the degree of autocorrelation. As mentioned in the chapter on the baseline, there are two alternatives for transforming data, first differencing or moving average.

IMPORTANT NOTE: If you transform data in one phase, you *must* use the same method of transformation, either moving average or differencing, to transform data in subsequent phases. Otherwise, an accurate comparison of phases, visually or statistically, is not possible.

When there is a trend in the data, the first difference method is preferred. To do this in *SSD for R*, use the following command:

>diffchart(cry, pcry, "B")

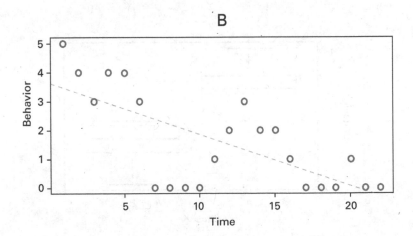

Figure 4.32 ABrf2() for crying behavior in the intervention phase

```
Console ~/ 
> ABrf2(cry,pcry,"B")
[1] "tf2="  "6.948"
[1] "rf2="  "1.019"
[1] "sig of rf2=" "0"
----------regression------------

Call:
lm(formula = A ~ x1)

Coefficients:
(Intercept)           x1
      3.649       -0.179

> 
```

Figure 4.33 Statistical output from ABrf2() for crying behavior

This function produces the graph displayed in Figure 4.34. As presented in this figure, the differencing does smooth out the data. The scores represent the degree of change between adjacent observations; thus, values can be positive, negative, or zero. In Figure 4.34, the dotted line represents the degree of difference between adjacent observations. The solid line represents the original data. Notice that the shape of both lines is very similar; removing the trend and autocorrelation in the dotted line, there is more confidence in what we observe. Because this appears to have possibly taken care of the autocorrelation problem, you will want to save the transformed data by responding appropriately to the prompt in the Console.

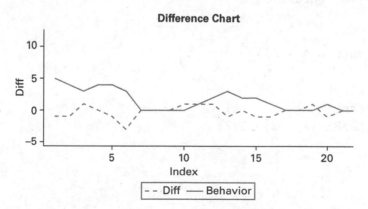

Figure 4.34 First order differencing for crying behvaior in the intervention phase

After you have saved the transformed data, use the **Getcsv()** function to open it. Once the dataset is open, type **attach(ssd)**. Once the dataset is attached, you can test it for autocorrelation by entering the following command:

>**ABrf2(diff, phase,"B")**

The statistical output of this is shown in Figure 4.35.

As we can see, differencing reduced the autocorrelation from an r_{f2} of 1.019 to an r_{f2} value of 0.209; additionally, what autocorrelation exists is no longer significant (sig of r_{f2} = 0.434). The slope is now 0.02597, indicating that the trend has also been removed.

Figure 4.36 (see p. 87) displays the graphical output from the function you entered. Notice that the regression line is flat, providing visual evidence that the trend was removed.

In the chapter on analyzing the baseline, you used the first difference to transform your baseline data. Now, both the phases have been transformed, and the autocorrelation has successfully been removed.

Another option for transforming data is with the **ABma()** function. As stated in Chapter 3, it is an appropriate function to use in trying to transform autocorrelated data when the data in a phase is highly variable.

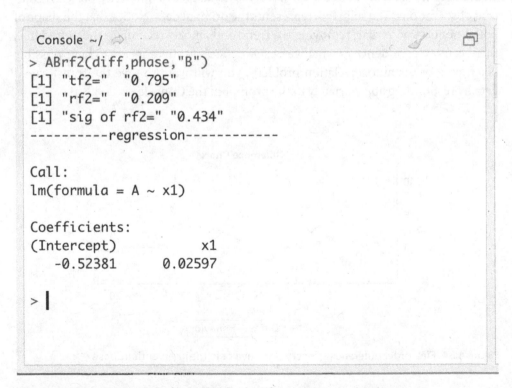

```
Console ~/
> ABrf2(diff,phase,"B")
[1] "tf2="   "0.795"
[1] "rf2="   "0.209"
[1] "sig of rf2=" "0.434"
----------regression------------

Call:
lm(formula = A ~ x1)

Coefficients:
(Intercept)              x1
   -0.52381        0.02597

>
```

Figure 4.35 Statistical results for **ABrf2()** for transformed intervention data

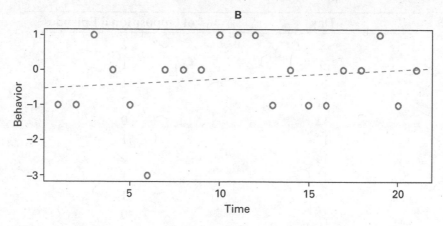

Figure 4.36 Graphical output from ABrf2() function of transformed data

In the next chapter on statistical significance, you will learn how to use the **Append()** command in *SSD for R* to combine transformed datasets in multiple phases (e.g., baseline and intervention) when testing for statistical significance.

Conclusion

In this chapter, we discussed how you might use descriptive statistics to compare two phases in a single-subject study. These include measures of central tendency, variation, trend, and effect size. Using Jenny as an example, you learned the commands necessary to calculate and interpret these statistics in *SSD for R*. Additionally, these commands generate supporting graphs, including box plots, SD band graphs, and line charts showing the mean, median, and the trimmed mean for any phase that you can use to compare any two phases.

SSD for R also provides three standard methods for computing effect size, which were discussed in detail. Additionally, four methods of evaluating effect size using non-overlap methods were examined. The chapter concluded with a discussion of autocorrelation in the intervention phase and how to consider dealing with this issue in light of the possibility of autocorrelation in a comparative (e.g., baseline) phase.

Chapter Exercises

Assignment 4.1—Working With Two Phases of Data

For this assignment, you will modify the spreadsheet you created in Chapter 1 with Brenda's baseline data. You will add data to continue tracking her behavior into the intervention phase. Add the following data to your EXISTING file:

Day	# of Oppositional Episodes
8	2
9	1
10	1
11	2
12	0
13	1
14	0
15	1
16	1
17	0
18	0
19	0
20	0
21	0
22	0
23	0

Once you add these data to your Google Sheet, do the following:

1. Import it into *RStudio* with a different file name than you used before.
2. Create a line graph labeling it appropriately. In the main title, include YOUR initials. For example, my main title could read, "Brenda's Oppositional Behavior—WZ"
3. Annotate the line graph to put a vertical line between phases.
4. Label the baseline phase "A" and the intervention phase "B"
5. Export the graph to a Word document and add your answer to the following questions: Based on the line graph alone, do you think that the social worker's intervention with Brenda is having the desired effect? Why or why not?

Assignment 4.2—Visual Comparison of Two Phases

For this assignment, you will continue to use the file you created in assignment 4.1 with baseline and intervention data for Brenda.

1. Re-create the graph you produced for the previous assignment by creating and labeling a line graph. Make the main title contain your own initials.
2. Add a vertical line between phases and label each phase as "A" or "B."
3. Add a mean line for each phase and label the line with the mean value for each phase.

4. Produce descriptive statistics for each phase by using the **ABdescrip**() function.

Then report the following:

1. Include the line graph in your report.
2. Analyze **each** phase of the data by reporting the phase, number of observations in the phase, the mean, sd, median, and range of data in the phases.
3. Visually compare phases: Does the problem appear to be accelerating, decelerating, or zero-celerating? What makes you think this?
4. Based on this analysis, if you were working with Brenda, what would you do at this point?

Assignment 4.3—Effect Sizes

1. Create a line graph of Brenda's tantrums, separating the phases with a vertical line. Take an educated guess about whether your intervention is "not effective," is "debatably effective," "moderately effective," or "very effective."
2. Consider the qualities of Brenda's tantrums (i.e., trending and autocorrelation in each phase) and determine which type(s) of effect sizes would be appropriate to use.
3. Calculate those effect sizes. Were you right in your visual analysis? Why do you think you might have been right or wrong?
4. Now, take what you have learned and make a practice decision. What would you do if Brenda was your client?

5

Statistical Tests of Type I Error

Introduction

In this chapter, we discuss tests that can be used to compare your data across phases. These are all tests of statistical significance that are used to determine whether observed outcomes are likely the result of an intervention or, more likely, the result of chance. Used correctly, statistical tests can help you make sounder decisions about your data (Bloom et al., 2009).

Usually, the first step in significance testing is to form a hypothesis of no difference, referred to as the null hypothesis. In single-system design, the typical null hypothesis, described as H_0, is that the target behavior did not change after the introduction of the intervention. For example, we could state the null hypothesis that Jenny's self-esteem did not change after participation in a social skills group. An alternative hypothesis, denoted as H_1 or H_A, is that the target behavior changed after the introduction of the intervention. This is a nondirectional hypothesis because, stated as such, change can be positive (i.e., desired) or negative (i.e., undesired). Continuing the Jenny example, we could test the hypothesis that Jenny's self-estemm did change after participation in the school's social skills group. A directional null hypothesis would be that the target behavior was either the same or worse after the intervention, such as Jenny's self-esteem not improving or deteriorating after participating in social skills. The alternative directional hypothesis would be that the behavior improved after the intervention, and we would see an improvement in Jenny's self-esteem after participating in the social skills group. We recommend the use of a nondirectional hypothesis because it allows you to establish if the change is positive or negative.

The purpose of a statistical test is to determine how likely it is that we are making an incorrect decision by rejecting the null hypothesis and accepting the alternative one. In statistics, this is referred to as a Type I error. In more common parlance, this is a "false positive." By using probability, tests of statistical significance quantify the likelihood of making a Type I error. Typically, in the social sciences, we are willing to accept the alternative hypothesis if the chance of making a Type I error is 5%, or a probability of .05, or less. This is typically shown as output in statistical programs as a p value (e.g., $p \leq .05$) or a sig value (e.g., $sig \leq .05$).

Statistical Significance Versus Practical Significance

It should be noted that if you observe statistical significance between phases, this does not guarantee that the observed differences will be noticeable in a

SSD for R. Charles Auerbach and Wendy Zeitlin, Oxford University Press. © Oxford University Press 2022.
DOI: 10.1093/oso/9780197582756.003.0006

real-life or practical setting. It is possible to compute statistically significant differences, but observe a small magnitude of change that is, perhaps, imperceptible to you or your client. Conversely, you could observe moderate or large clinically meaningful differences between the baseline and intervention phases that are not statistically significant due to the way those computations are calculated.

Therefore, we recommend, and best practice dictates, that if you are going to conduct hypothesis tests, you should holistically consider both the results of your hypothesis test and effect size calculations when making practice decisions.

Consider the following example: You compare phases with a hypothesis test, but the difference between phases is not statistically significant. This could lead you to the conclusion that there is no difference in the client's behavior after the intervention, and you decide to discontinue the intervention. If, however, you continue your analysis, you may notice that there is a moderate change in the desired direction, it is possible that your intervention is having the desired impact, and more time and data are needed with the current intervention to observe statistically significant findings.

A number of tests of significance are presented in this chapter: statistical process charts (SPCs), proportion/frequency, chi-square, the conservative dual criteria (CDC), robust conservative dual criteria, the *t* test, and analysis of variance (ANOVA). How and when to use each of these tests is also discussed.

Statistical Process Control Charts

As Orme and Cox (2001) pointed out, SPC charts, also known as Shewhart charts or control charts, are very familiar to social work research. This section is based on the work of Orme and Cox (2001), who did an excellent job of describing the use of SPC charts, which come out of the discipline of industrial quality control (Orme & Cox, 2001). A number of these charts were described at length by Bloom et al. (2009), and, in recent years, these charts have become popular in healthcare quality improvement, which requires making changes in processes of care and service delivery. In the social and health sciences, SPC charts are used to measure if process changes are having a desired beneficial effect (Benneyan, Lloyd, & Plsek, 2003; Mohammed & Worthington, 2012; Polit & Chaboyer, 2012; Smith et al., 2012; Tasdemir, 2012; Woodall, 2006).

Since a stable process will follow a normal distribution, we expect approximately 95% of collected data to fall within two standard deviations and approximately 99% to be within three standard deviations (Benneyan et al., 2003). The probability that any one observation would fall above or below three standard deviations is .0027, which would be considered statistically significant because it is less than .05; however, it should be noted that the standard deviation for SPC charts is derived differently from the form we discussed in the section on variation.

In the examples we provide in this chapter, we use the more conservative three standard deviation band approach in examining SPC charts; however, two standard deviation bands have been used in prior research and are also considered acceptable. Using two standard deviations, as opposed to three, reduces the chances of making a Type II error, or a false negative, but increases the chances of making a Type I error (0.025 compared to 0.0027 for a two-tailed test). In *SSD for R*, you have the option of choosing SPC charts with one, two, or three standard deviations.

With the enactment of the Patient Protection and Affordable Care Act (2010), there has been an increased need for accountability. *SSD for R* produces a number of SPC charts that can be used specifically for this purpose.

The SPC charts have a number of advantages in single-system data analysis. Polit and Chaboyer (2012) described a number of these. First, they can be used with data with a variety of characteristics. They can be grouped by a time-based variable such as days or weeks. Additionally, the impact of an intervention can be detected even as data are being collected. Because SPC charts are easy to understand, they are an excellent form of communicating change over time to untrained audiences (Polit & Chaboyer, 2012).

There are a number of SPC charts presented in this chapter, and not all charts are appropriate for every situation. Decision trees are available in Appendix C to help you select the chart(s) that are appropriate for your data and reporting needs (Orme & Cox, 2001).

All SPC charts presented in this section could be impacted by autocorrelation, so it is always a good idea to test for it prior to generating any of these. According to Wheeler (2004), SPC charts work well when autocorrelation is modest (Wheeler, 2004). The adequacy of these procedures is diminished when the lag-1 autocorrelation is high (i.e., greater than or equal to 0.6).

X-bar Range Chart (X̄-R Chart)

The X-bar range chart (X̄-R chart) can be used when there are multiple observations per sample (Orme & Cox, 2001). For example, in hospitals, social work services are often extended to the emergency department (ED) to decrease the rate of unnecessary hospital admissions due to nonmedical-related factors. The director of social work could track the daily percentage of admissions for nonmedical reasons. A retrospective baseline could, for example, be acquired for 6 weeks prior to the implementation of social work services and 6 weeks postintervention. If data were collected on a daily basis, there would be 12 samples (6 weeks of baseline plus 6 weeks of intervention), with seven observations (days) per sample. Using the mean of these samples increases the accuracy of our measure (Mohammed & Worthington, 2012; Orme & Cox, 2001).

Figure 5.1 displays the X̄-R chart for this example of ED data. To begin creating this chart yourself, you will need to use data from the *ed.csv* dataset by entering the following command in the Console:

>**Getcsv()**

Once the dataset is open, attach it by entering the following command in the Console:

>**attach(ssd)**

To begin, it is important to consider each variable in the dataset. Note that when you open this dataset you can view the contents by clicking on the spreadsheet icon to the right of the file in the Environment pane. When you do this, you will note that there are three variables listed: *admits*, *admitweek*, and *padmit*. The *admits* variable contains the data on the percentage of nonmedical admissions for each day, *admitweek* is the grouping variable for the eight samples (i.e., weeks), and *padmit* is the phase variable with "A" representing the baseline and "B" representing the intervention.

To create the annotated X-R chart, enter the following *SSD for R* commands in the Console. After each annotation, you will be prompted to accept the alteration to the plot. Once you accept each one by entering *y* in the Console, note that you will not be able to change that annotation again.

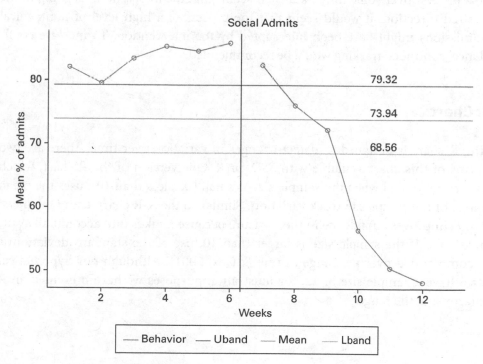

Figure 5.1 Example of X-bar R chart

Command	Purpose
XRchart(admits,admitweek, 3, "weeks","mean % of admits","Social Admits")	Creates SPC chart using three standard deviations
ABlines(admits)	Adds line between phases
ABtext("73.94")	Enters mean text
ABtext("79.32")	Enters Uband text
ABtext("68.56")	Enters Lband text
SPClegend()	Creates a legend

Note that once you create a legend, you will no longer be able to alter the graph. Therefore, we recommend that you add this as a last step in creating SPC charts. Be sure that your graph looks exactly as you would like it before adding the legend.

In the Console, you will see the values for the upper band (Uband), mean, and lower band (Lband). These values were used to label the $\bar{X}$-R chart.

In this case, the desired zone would be below the lower band (68.56) because the hospital's goal is to reduce the number of nonmedical admissions. The last 3 weeks of the intervention are in the desired zone, showing that the intervention has some promise. The hospital administrator would probably want to continue measuring this process over some time because it seems that the intervention is going in the desired direction. It would seem that the process of a high level of nonmedical admissions might have been interrupted by the intervention. To increase confidence, continued tracking would be recommended.

R Chart

The R chart is designed to detect changes in variation over time. There are two forms of this chart available with *SSD for R*. One version of the R chart, which is recommended when the sample size is small (i.e., less than 10), uses the mean range of the samples to track variation. Similar to the $\bar{X}$-R chart, using the mean range improves confidence in the measure because it takes into account all available data. If the sample size is larger than 10, use of the standard deviation is recommended over the range (Orme & Cox, 2001). Although our hypothetical data have a sample size of 12, for illustration purposes we have presented an R chart using the range.

Figure 5.2 was created with data from the *ed.csv* file by entering the following commands in the Console:

Command	Purpose
Rchart(admits, admitweek, 3,"weeks","mean range of admits","Social Admits")	Creates SPC chart using three standard deviations
SPCline()	Adds line between phases; click twice to indicate the top and bottom of the line
ABtext("24.69")	Enters mean text
ABtext("12.83")	Enters Uband text
ABtext("0.97")	Enters Lband text
SPClegend()	Creates a legend

Figure 5.2 shows that all sample mean ranges for the 12 weeks are within the two bands, indicating that the variation for this process remained stable.

The commands in the following table can be used to create an R chart using the standard deviation as the measure of process:

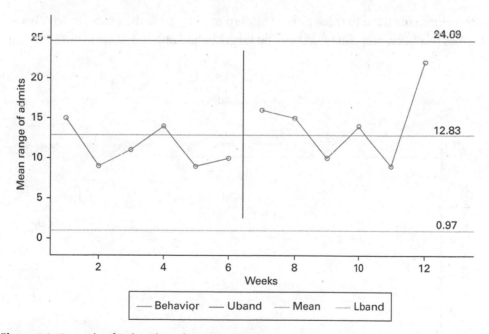

Figure 5.2 Example of R chart based on mean range

Command	Purpose
Rchartsd(admits, admitweek, 3, "weeks", "mean SD of admits","Social Admits")	Creates SPC chart using three standard deviations
SPCline()	Adds line between phases; click twice to indicate the top and bottom of the line
ABtext("8.61")	Enters Uband text
ABtext("4.48")	Enters mean text
ABtext("0.34")	Enters Lband text
SPClegend()	Creates a legend

Figure 5.3 shows virtually the same trend displayed in Figure 5.2. More importantly, all the weekly standard deviations are within the two bands, demonstrating that the variation for this process remained stable. As we described previously, extreme variation influences the stability of a measure.

Proportion Chart

Often clinicians need to track a client system to determine the extent to which a set of tasks has been completed. When the target behavior has a binary outcome (e.g.,

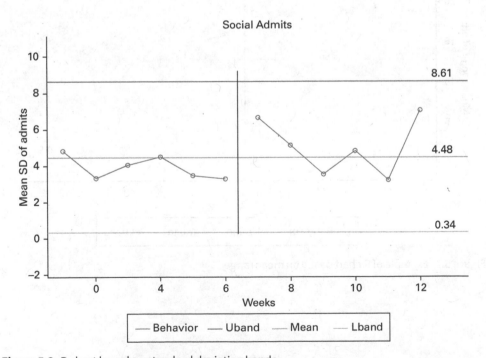

Figure 5.3 R chart based on standard deviation bands

completing or not completing a task), comparing the proportion of task completion over time or between phases can be measured. Attendance (vs. nonattendance) in a group, child welfare workers completing records on time (vs. not completing on time), workers completing home visits (vs. not completing home visits), whether a patient is compliant with taking medication (vs. noncompliant), and if a patient with hearing loss wears their hearing aid (vs. not wearing the hearing aid) are all examples of task completion that can be assessed using proportion charts (P charts). For a more in-depth discussion of this procedure, we suggest you refer to J. G. Orme and Cox (2001) and Mitra (2008).

As an example, open the *Jennyab.csv* dataset, attach it, and view the variables by clicking on the spreadsheet icon to the right of the *ssd* file in the Environment pane. In this example, we use the following three variables: *group*, *wgroup*, and *pgroup*. The *group* variable is a binary measure of group attendance, with "0" indicating that Jenny did not attend the daily group and "1" indicating that she did attend. The variable *wgroup* is the week of attendance, and *pgroup* is the phase variable indicating whether the measure was obtained during the baseline or intervention. To improve attendance, after the fifth week an intervention was introduced; children who had perfect weekly attendance were given an extra 20 minutes of free play on Friday afternoons.

These data have 15 samples (i.e., weeks) and five observations (i.e., days) per sample. Entering the following commands will create the p chart in Figure 5.4.

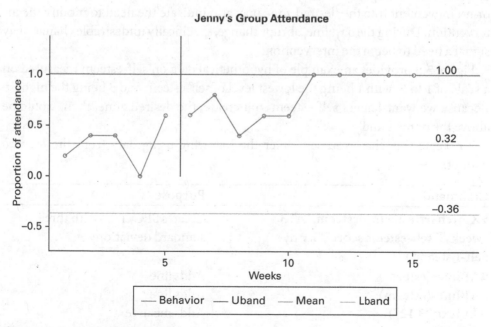

Figure 5.4 P chart based on 3-sd bands

Command	Purpose
Pchart(group,wgroup, 3, "weeks", "proportion of attendance", "Jenny's Group Attendance")	Creates SPC chart with three standard deviations
ABlines(group)	Adds line between phases
ABtext("0.32")	Enters mean text
ABtext("1.00")	Enters Uband text
ABtext("−0.36")	Enters Lband text
SPClegend()	Creates a legend

Figure 5.4 shows that all the samples during the baseline are within the upper and lower bands. After the intervention, the proportion of attendance increases, and, during the last 5 weeks, Jenny attended all group sessions. The findings from the chart would provide evidence to continue the intervention.

X-Moving-Range Chart

Like the previous SPC charts, the X-moving-range chart (X-mR chart) can be used to detect changes within and between phases. This chart should not, however, be used when there is a trend in the data. It can be used, though, when you are not using samples of data, but individual data points, such as what we have seen with the majority of Jenny's data. As with the previous charts, large unexpected changes in the undesired zone may indicate the need to modify the intervention. Additionally, slow or no movement into the desired zone may also indicate the need to modify the intervention. During the baseline, abrupt change, specifically undesirable change, may signal a need to begin the intervention.

Figure 5.5 provides an example of hypothetical data on self-esteem measured on a scale of 1 to 5, with 1 being the lowest level of self-esteem and 5 being the highest. Because we want Jenny's self-esteem to increase, the desired zone, then, would be above the upper band.

To create this chart yourself, enter the following using data from the *Jennyab* dataset:

Command	Purpose
>Xmrchart(esteem, pesteem, "A", 3, "weeks","self-esteem score","Jenny's Self-Esteem")	Creates SPC chart with three standard deviations
>ABlines(esteem)	Adds line
>ABlines(esteem)	Adds line
>ABtext("3.12")	Adds mean text

>ABtext("5")	Adds upper band text
>ABtext("2")	Adds lower band text
>ABtext("A")	Adds text
>ABtext("B")	Adds text
>ABtext(expression(A[1]))	Adds text with subscript
>SPClegend()	Adds legend

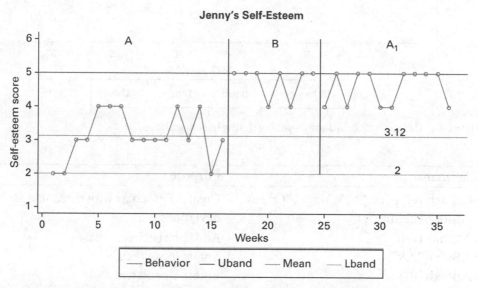

Figure 5.5 X-mR chart based on Jenny's self-esteem during baseline phase

Note that, in this example, we have three phases: a baseline (A), an intervention (B), and a return to baseline (A₁), as the intervention was removed after the 24th week. In this case, the intervention is to encourage Jenny's parents to be more responsive to her at home. After the intervention, Jenny's level of self-esteem increased and was sustained when the intervention was removed (A₁).

C Chart

It is suggested that when the outcome is a count of behavior the C chart should be utilized. Examples of behaviors that are counts would be the number of times a child displays aggressive behavior, the number of times a child leaves his or her seat, or the number of times a child yells out an answer.

As an example, you can re-create Figure 5.6 (see p. 100) by entering the following commands in the Console using the *Jennyab* dataset:

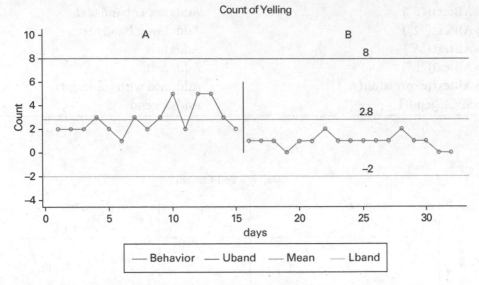

Figure 5.6 C chart based on Jenny's yelling during baseline phase

Command	Purpose
>Cchart(yell,pyell,"A", 3,"days", "Count", "Count of Yelling")	Creates SPC chart with three standard deviations
>ABlines(yell)	Adds line between phases
>ABtext("2.8")	Enters mean text
>ABtext("8")	Enters Uband text
>ABtext("–2")	Enters Lband text
>ABtext("A")	Enters text for the A phase
>ABtext("B")	Enters text for the B phase
>SPClegend()	Creates a legend for the graph

As Figure 5.6 illustrates, all the points are within the upper and lower bands. Although there are no unusual observations during either the baseline or the intervention, the yelling was consistently lower after the intervention was initiated.

Additional Notes on SPC Charts

Once again, all the charts presented in our examples used three standard deviation bands. Bloom et al. (2009) pointed out that using the bands with three standard deviations creates a condition where the probability of making a Type I error is low (0.0027); therefore, they posited that the X-mR chart will be reliable even under conditions of autocorrelation.

On the other hand, Mohammad and Worthington (2012) suggested attribute charts, like the p chart and C chart, can at times be misleading. Therefore, they suggested that using more than one SPC chart can reduce flawed interpretations of findings. Under conditions of normal variation, the X-mR chart will be in close agreement with the p chart and C chart. If this is not the case, it could signal some underlying problem that could lead to misinterpretation (Mohammed & Worthington, 2012).

Other Tests of Type I Error

In addition to the numerous SPC charts described in the preceding section, there are other tests of Type I error that are appropriate for use in single-system research. *SSD for R* has functions to conduct these tests; however, the decision to use a particular test of Type I should be based on the nature of the data in each phase. See Appendix C for decision trees to help you select the most appropriate test to use.

Proportion/Frequency or Binomial Test

The proportion/frequency procedure compares typical patterns between the baseline and the intervention. In this statistical test, the binomial distribution is used to test for statistical significance between phases (Bloom et al., 2009).

Based on the properties of the normal distribution, we know that 68.2% of all scores should fall within one standard deviation of the mean. We can, therefore, use this knowledge to define desired and undesired zones. If higher values signify improvement, then the desired zone will be above the upper band. On the other hand, if lower values indicate improvement, the desired zone will be below the lower band. Two standard deviations, which encompass 95% of scores in a normal distribution, could also be used to define the desired zone; however, the probability of a score falling above or below two standard deviations is only 0.025.

Let's use Jenny's yelling behavior as an example. We begin by constructing a standard deviation band graph that extends the mean and standard deviation bands calculated for the baseline into the intervention phase. In this case, the desired zone would be below the lower band because we want the instances of yelling to decrease. Any other location on the graph would be considered undesirable. Figure 5.7 (see p. 102) displays this graph. The following commands with the *Jennyab* file open produced the figure:

Command	Purpose
SD1(yell,pyell,"A","days", "Count", "Count of Yelling")	Creates graph
ABlines(yell)	Adds line between phases
ABtext("+1sd = 4.06")	Enters text between quotes
ABtext("–1sd=1.54")	Enters text between quotes
ABtext("mean=2.8")	Enters text between quotes

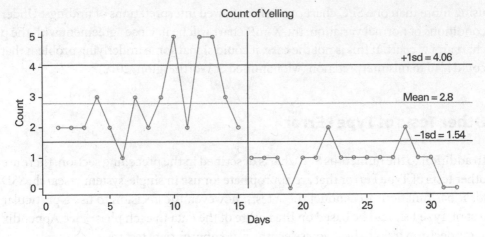

Figure 5.7 Line graph of Jenny's yelling with 1-sd bands based on baseline phase

As the figure illustrates, many more of the observations of the target behavior are in the desired zone in the intervention phase, which is below the lower standard deviation band, compared to the baseline. This shows that the yelling behavior dropped sharply during the intervention.

We can test if the proportion of scores in the desired zone during the intervention was significantly different from the baseline using the **ABbinomial()** function. The first step to doing this is to count the number of observations in the desired zone for both the baseline and intervention, which are 1 and 15, respectively.

The following command entered in the Console will produce the output in Figure 5.8:

>**ABbinomial(pyell, "A", "B", 1, 15)**

```
> ABbinomial(pyell, "A", "B", 1, 15)

        Exact binomial test

data:  successB and tmaxB
number of successes = 15, number of trials =
17, p-value = 2.73e-16
alternative hypothesis: true probability of success is not equal to 0.06666667
95 percent confidence interval:
 0.6355908 0.9854207
sample estimates:
probability of success
           0.8823529
```

Figure 5.8 Console output for binomial test of Jenny's yelling across phases

Let's dissect this function: Here *pyell* is the phase variable for Jenny's yelling behavior; "*A*" is the baseline phase or the phase that is to serve as the basis for comparison, "*B*" is the intervention phase, *1* denotes the number of observations in the desired zone for the baseline phase, and *15* is the number of observations in the desired zone during the intervention. With the **ABbinomial()** function, any two phases can be compared, but the letter entered after the phase variable needs to be the phase to which the second phase is compared.

The null hypothesis is that there is no difference between the rates of success, as defined by the number of points in the desired zone, in the two phases. The alternative hypothesis is that the rates are different. The *p* value in Figure 5.8 is displayed in scientific notation, denoting that it is very low. If you want to view the entire decimal value, there are two options. The first turns off scientific notation for your entire *R* session. Enter the following in the Console:

```
options(scipen=999)
```

Alternatively, you can use the following **SSDforR** function:

```
>SN(2.73e-16)
```

As we discussed in the beginning of this chapter, we need a significance value of .05 or less to reject the null hypotheses and therefore accept the alternative. Because the *p* value, or significance level, is less than .05 we can reject the null. Because 15 of 17 observations during the intervention were in the desired zone during the intervention, the rate of success is 0.8823529. This simply means that 88% of the observations were in the desired zone during the intervention. We can compare this to the rate of success during the baseline, which is listed next to *true probability of success is not equal to* in the Console. In this case, only 1 observation out of 15, or 6.7%, was in the desired zone during the baseline.

Another Example

Proportion/frequency can also be used to help track task completion and can be used as an alternative to SPC p charts. When the target behavior has a binary outcome (i.e., completing or not completing a task), comparing the proportion of tasks completed over time or between phases can be measured. One example is a client's attendance in a group (attendance vs. nonattendance).

Let's use Jenny's group attendance as an example. The behavior variable group is coded as "1" for attended and "0" for did not attend. We can use the *R* **table()** function to compare the frequency of attendance during the baseline to the intervention. To do this, enter the following in the Console:

```
>table(group, pgroup)
```

```
> table(group, pgroup)
          pgroup
group   A   B
    0  17  10
    1   8  40
```

Figure 5.9 Console output of **table()** function

Figure 5.9 is displayed in the Console.

From the output in the table in Figure 5.9, we can see that Jenny attended the group 8 out of 25 (17 + 8) sessions during the baseline compared to 40 of 50 (10 + 40) sessions during the intervention.

The **ABbinomial()** function can now be utilized to test if the rate of attendance during the intervention was statistically different from the baseline. In this example, we would enter the following into the Console with the output displayed in Figure 5.10:

>**ABbinomial(pgroup,"A","B", 8, 40)**
>**SN(3.934e-12)**

The significance value is less than .05, and the null hypothesis can therefore be rejected. The output shows that there was an increase from a 32% (0.32) attendance success rate during the baseline to 80% (.8) during the intervention.

Using Proportion/Frequency With Client Goals

Recall that Gloria is experiencing depression, and you would like to assess whether her sleeping improves with the introduction of cognitive behavioral therapy (CBT). Since clients with depression can sleep too much or too little, you work with Gloria to determine a healthy range of sleep, between 6 and 8 hours per night.

You can use the **Gline()** function to create boundaries around your desired zone in a line graph. Begin by opening and attaching the *GloriaABC* file.

```
> ABbinomial(pgroup, "A", "B", 8, 40)


        Exact binomial test

data:  successB and tmaxB
number of successes = 40, number of trials = 50, p-value = 3.934e-12
alternative hypothesis: true probability of success is not equal to 0.32
95 percent confidence interval:
 0.6628169 0.8996978
sample estimates:
probability of success
                  0.8
```

Figure 5.10 Console output for binomial test of Jenny's group attendance across phases

> ABplot(sleep, psleep, "night", "hours","Gloria's sleeping")

You will want to separate your phases with a vertical line:

> ABlines(sleep)

Then create your desired zones with the addition of **Gline()**, specifying lines at both 6 and 8 hours of sleep. Follow the prompts in the Console until your plot looks like the one illustrated in Figure 5.11.

Notice that by using the **Gline()** function you are able to customize what is considered desirable. This function, then, can be used in collaboration with work with clients to help assess the degree to which they are reaching goals that are meaningful to them.

We can now count two desirable values in the baseline and six in the intervention. With this in mind, we can run the **ABbinomial()** function:

> ABbinomial(psleep, "A", "B", 2, 6)

The output found in the Console is illustrated in Figure 5.12 (see p. 106).

The output illustrates that Gloria slept a healthy amount 40% of the time during the baseline and 6 out of 10 times, or 60% of the time, during the intervention. While we see an improvement, the p value of .2126 is greater than .05, so we are unable to reject the null hypothesis.

This is disappointing, but remember that the number of observations in each phase impacts p values, so do not give up just yet. Perhaps with more time, statistical significance could be achieved.

Chi-Square

The chi-square (χ^2) is a widely used statistic in the helping professions. The chi-square is a nonparametric statistic used to test for Type I error when data are categorical. Because it is a nonparametric test, we assume independence; therefore, the chi-square should not be used when autocorrelation is problematic. Before deciding

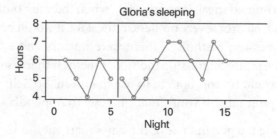

Figure 5.11 Line graph of Gloria's sleeping with defined goal lines

```
> ABbinomial(psleep, "A", "B", 2, 6)

        Exact binomial test

data:  successB and tmaxB
number of successes = 6, number of trials = 10, p-value = 0.2126
alternative hypothesis: true probability of success is not equal to 0.4
95 percent confidence interval:
 0.2623781 0.8784477
sample estimates:
probability of success
                  0.6
```

Figure 5.12 Console output for binomial test of Gloria's sleep across phases

to use chi-square or any other test for Type I error, you should test both phases to see if autocorrelation is a problem.

In this example, let's examine Jenny's yelling behavior more closely. You can test two phases using the **ABrf2()** function to test for autocorrelation of Jenny's yelling behavior, by entering the following commands in the Console, the results of which are displayed in Figure 5.13 (see p. 107):

>**ABrf2(yell, pyell, "A")**

>**ABrf2(yell, pyell, "B")**

The r_{f2} values in Figure 5.13 display small nonsignificant rf_2 values of 0.267 and 0.383, respectively, for the baseline and intervention phases. Because we met the assumption that the data in both phases are sufficiently independent, we can proceed with a chi-square test.

Because the chi-square is used for categorical data, a method to create frequencies is necessary. Creating a desired zone similar to what was done for the proportion/frequency can be used. Five methods for doing this are available in *SSD for R*: the median, the mean, the trimmed mean, the ordinary least squares (OLS) regression line, and the robust regression line. You should use whichever method is most appropriate for your situation.

The median or trimmed mean is appropriate when there are outliers and no trend. If there were a trend, but not severe outliers in the data, it would be more appropriate to use the OLS regression method. If there were outliers and a trend, it would be preferable to use the robust regression method. Finally, if there is no trend and there are no outliers, it would be appropriate to use the mean. To help you choose which method is most appropriate for your situation, refer to Appendix C.

To better illustrate the general use of the chi-square methods, we can look more closely at Jenny's yelling behavior, the results of which are displayed in Figures 5.14

```
> ABrf2(yell, pyell, "A")
[1] "tf2="  "0.801"
[1] "rf2="  "0.267"
[1] "sig of rf2=" "0.432"
----------regression-----------

Call:
lm(formula = A ~ x1)

Coefficients:
(Intercept)            x1
     1.7429        0.1321

> ABrf2(yell, pyell, "B")
[1] "tf2="  "1.294"
[1] "rf2="  "0.383"
[1] "sig of rf2=" "0.209"
----------regression-----------

Call:
lm(formula = A ~ x1)

Coefficients:
(Intercept)            x1
    1.11765       -0.01961
```

Figure 5.13 Console output assessing autocorrelation of yelling in two phases

(see p. 108) and 5.15 (see p. 109). This example uses the baseline mean to form the desired zone. We chose this method because there are no outliers in the data, and there is not a significant trend in either phase. Because we want the number of yelling episodes to decrease, the desired zone is below the regression line. Enter the following command in the Console:

>**meanbelow(yell, pyell,"A","B")**

The first part of the output in Figure 5.14 displays the frequencies and percentages. The "[,1]" is the baseline phase, and "[,2]" is the intervention phase. "FALSE" is above the reference line and represents what is not desirable, while "TRUE" is below the mean line, or in the desired zone. For the baseline, eight observations (53.3%) are in the desired zone, and seven (46.7%) are in the undesired zone. For the intervention, all 17 observations (100%) are in the desired zone.

The next section shows the frequencies are the percentages for all observations across all phases. All of the undesired scores (100%) occurred in the baseline. Over two thirds (68%) of the observations in the desired zone occurred in the intervention.

The p value for the chi-square is .001439. Because the p value is less than .05, the null hypothesis can be rejected. Because the entire table is 2×2 (two columns by two rows), Fisher's exact test is calculated. This test provides a more precise calculation of significance in small tables with small cell sizes. In this case, the p value = .001912. Based on this result, we would conclude that there was a statistically significant

```
> meanbelow(yell, pyell, "A", "B")
       [,1] [,2]
FALSE   7    0
TRUE    8   17
       [,1] [,2]
FALSE 100    0
TRUE   32   68
          [,1] [,2]
FALSE 46.66667    0
TRUE  53.33333  100


        Pearson's Chi-squared test

data:  ctbl
X-squared = 10.155, df = 1, p-value = 0.001439



        Fisher's Exact Test for Count Data

data:  ctbl
p-value = 0.001912
alternative hypothesis: true odds ratio is not equal to 1
95 percent confidence interval:
 2.263412       Inf
sample estimates:
odds ratio
        Inf

Warning message:
In chisq.test(ctbl, correct = FALSE) :
   Chi-squared approximation may be incorrect
```

Figure 5.14 Console output of chi-square test comparing Jenny's yelling across phases

higher frequency of scores in the desired zone during the intervention compared to the baseline.

The plot shown in Figure 5.15 (see p. 109) illustrates both baseline and intervention phases with the horizontal line placed at the mean for the baseline, denoting the boundary for the desired zone.

If the desired zone were above the mean line, the command would be as follows: **meanabove (yell,pyell,"A","B")**. If the median was to be used to determine the desired zone, the commands would be **medbelow(yell,pyell, "A","B")** or **medabove(yell,pyell,"A","B")**. If the OLS regression line was used, the commands would be **regbelow(yell,pyell,"A","B")** or **regabove(yell,pyell, "A","B")**. If the trimmed mean was to be used, the commands would be **trim dbelow(yell,pyell,"A","B")** or **trimabove(yell,pyell,"A","B")**. If you were using

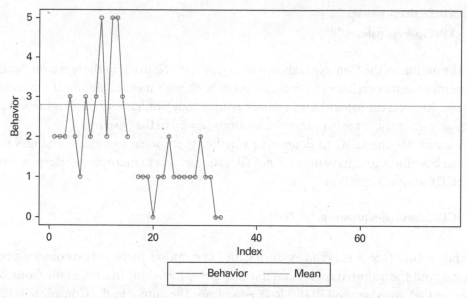

Figure 5.15 Visual output of chi-square test comparing Jenny's yelling across phases

robust regression and the desired zone was above the line, the command would be **robregabove(yell, pyell, "A", "B")**. If you were using robust regression and the desired zone was below the line, the command would be **robregbelow(yell, pyell, "A", "B")**.

Conservative Dual Criteria

The CDC is a test of Type 1 error that works well when data have relatively high autocorrelation provided there are between 5 and 20 intervention data points (Fisher et al., 2003; Morgan, 2008; Stewart et al., 2007; Swoboda et al., 2010; Tate & Perdices, 2019). The CDC uses two lines to define a desired zone: the mean and the regression line of the comparison phase, usually the baseline. If the preferred outcome is an increase in the target behavior, then the desired zone would be above both lines, and the **CDCabove()** function should be used. On the other hand, if lower scores are preferred, then the desired zone would be below both lines, and the **CDCbelow()** function should be utilized.

In our example of proportion/frequency, we considered whether there was a significant change in Gloria's sleeping prior to and after the introduction of CBT; however, that was based on a goal set between Gloria and her social worker. Let's consider whether there were any statistically significant differences in her sleeping from baseline to intervention. Begin by assessing the degree to which there is autocorrelation in both phases:

>**ABrf2(sleep, psleep, "A")**
>**ABrf2(sleep, psleep, "B")**

The output in the Console is shown in Figure 5.16. Notice that there is a moderate amount of autocorrelation in both phases even though it is not significant in either phase. To be on the safe side, we will not assume independence of observations, particularly since there are very few observations ($n = 5$) in the baseline.

We can use the CDC to determine whether there were significant changes between baseline and intervention. Since Gloria's goal was to increase her sleep, we use the **CDCabove()** function:

>**CDCabove(sleep, psleep, "A", "B")**

As Figure 5.17 (see p. 111) shows, during the intervention, there were six observations above both the adjusted regression line and the adjusted mean line. In the Console, you see that noted as the TRUE TRUE condition. The output in the Console tells you that eight observations are needed in this desired zone in order to achieve statistical

```
> View(ssd)
> ABrf2(sleep, psleep, "A")
[1] "tf2="    "-0.632"
[1] "rf2="    "-0.563"
[1] "sig of rf2=" "0.541"
----------regression------------

Call:
lm(formula = A ~ x1)

Coefficients:
(Intercept)           x1
        5.5         -0.1

> ABrf2(sleep, psleep, "B")
[1] "tf2="    "1.567"
[1] "rf2="    "0.664"
[1] "sig of rf2=" "0.138"
----------regression------------

Call:
lm(formula = A ~ x1)

Coefficients:
(Intercept)           x1
     4.8000       0.1818
```

Figure 5.16 Console output assessing autocorrelation of sleep in two phases

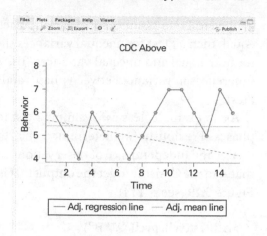

Figure 5.17 Output results of CDC test comparing sleep across phases

significance, and we are short by two. While things seem to be going in the right direction, this test reveals that there are no significant differences between the phases at this point.

There is a robust version of the CDC, **RobustCDCbelow()** and **RobustCDCabove()**, which replaces the mean with the trimmed mean and the OLS regression line with the robust regression line. Brossart, Parker, and Castillo (2011) made the case for the use of robust regression because it is less susceptible to the influence of outliers; therefore, it would be advisable to use a robust form of the CDC function if there are outliers in either phase.

It should be noted that, unlike some of the other tests you have read about in this section, there is no p value or significance value presented. You simply are provided the minimum number of data points that have to be in the desired zone in the intervention in order to achieve statistical significance.

t-test

The proportion/frequency test, the chi-square, and the CDC are based on comparing the number of observations in some desired zone in one phase to another. The t test compares the differences in the means between phases. The null hypothesis is that the difference in means between the phases is zero (i.e., there is no change between phases). If the t-test is statistically significant (i.e., $p \leq .05$), then we accept the alternative that the mean differences are greater than zero. The difference, however, could be improvement or deterioration in the behavior. If higher values are desired, the difference should be positive, and if lower values are desired, the difference should be negative.

There are some caveats for the use of the t test. It should not be used if there is a trend in the data or if either phase has problematic autocorrelation (Auerbach & Schudrich, 2013).

The *t* test also assumes equal variation between the phases. If the variances are unequal, then a *t* test for unequal variances should be used. *SSD for R* produces results for both equal and unequal variances. The **ABttest()** function allows you to test the differences in variance between phases and provides output for both versions of the *t* test.

As an example, let's use a simple *t* test to look at Jenny's yelling behavior across phases. Previously, we determined that there was no significant trend in either phase, and independence of observations can be assumed. Enter the following command in the Console to get the output that is displayed in Figure 5.18 (see p. 113) and Figure 5.19 (see p. 114):

>**ABttest(yell, pyell, "A", "B")**

There are three tests displayed in the Console panel to the left of the graphical output: a *t* test for equal variances (*Two Sample t-test*), a test for equality of variances (*F test to compare two variances*), and a *t* test for unequal variances (*Welch Two Sample t-test*). Because the *p* value for equality variances is less than .05 (.002407), we can reject the null hypothesis that the variances are not different and accept the alternative that they are not equal. Therefore, we use the *t*-test results for unequal variances to determine whether there are statistically significant differences between the baseline and the intervention. Looking at the results in this case, the null hypothesis is rejected because the *p* value is less than .05 (4.685e-05 or .00004685). Therefore, we accept the alternative that the mean difference between the phases is greater than zero. We also observe that the mean went down from 2.8 in the baseline to 0.9411765 in the intervention. We can therefore conclude that the introduction of the intervention was associated with an improvement in behavior.

The bar plot depicting this is shown in the Plots pane to the right of the Console.

Note that any two phases can be compared with *t* tests. If, for example, there were a second intervention (C), you could compare the mean of that phase against the baseline or against the first intervention (B), but not both. One-way ANOVA is a better choice if you want to compare more than two phases simultaneously.

One-Way Analysis of Variance

The one-way ANOVA is an extension of the *t* test. The assumptions are the same as required for the *t* test (i.e., no autocorrelation, no trend, equal variances, and normal distribution). However, this test is different because more than two phases can be compared simultaneously.

If we analyzed a behavior with three phases—a baseline (A), an intervention (B_1), and the introduction of a more intense intervention (B_2)—three separate *t* tests would be needed: A compared to B_1, B_1 compared to B_2, and A compared to B_2.

```
> ABttest(yell, pyell, "A", "B")

        Two Sample t-test

data:  A and B
t = 5.4965, df = 30, p-value =
5.735e-06
alternative hypothesis: true difference in means is not equal to 0
95 percent confidence interval:
 1.168160 2.549487
sample estimates:
mean of x mean of y
2.8000000 0.9411765


        F test to compare two variances

data:  A and B
F = 5.181, num df = 14, denom df =
16, p-value = 0.002407
alternative hypothesis: true ratio of variances is not equal to 1
95 percent confidence interval:
  1.839162 15.145963
sample estimates:
ratio of variances
        5.180952


        Welch Two Sample t-test

data:  A and B
t = 5.2611, df = 18.7, p-value =
4.685e-05
alternative hypothesis: true difference in means is not equal to 0
95 percent confidence interval:
 1.118519 2.599128
sample estimates:
mean of x mean of y
2.8000000 0.9411765
```

Figure 5.18 Console output results of t-test comparing yelling across phases

A one-way ANOVA offers a single test to determine whether the three phases differ from one another. The use of the ANOVA in this case, then, is advantageous over multiple *t* tests as multiple tests of significance are more likely to lead to spurious results.

Let's consider Gloria's case, in which her depression was measured prior to intervention, after the introduction of an intervention, and again when the intervention was intensified by doubling the number of sessions she was having each week.

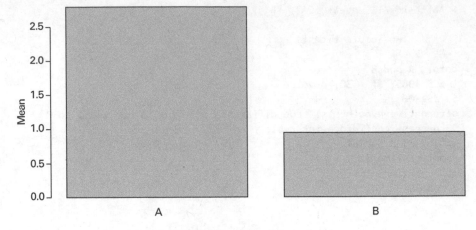

Figure 5.19 Visual comparison of mean yelling across phases

We can proceed with conducting the one-way ANOVA because there were no significant trends in any phase, and autocorrelation in each phase was low and not significant. As a refresher from Chapter 4, we can see the descriptive statistics for Gloria's depression by entering the following in the Console.

>**ABdescrip(depress, pdepress)**

Output displayed in the Console reminds us that Gloria's baseline mean depression was 37.8 (sd = 5.85). After the introduction of the intervention, her mean depression dropped to 24.8 (sd = 4.92), and when the intervention was intensified, her mean depression score dropped further to 18.0 (sd = 5.12).

With Gloria's dataset open and attached, type the following command in the Console:

>**ABanova(depress,pdepress)**

Figures 5.20 (see p. 115) and 5.21 (see p. 115) display the results of the one-way ANOVA.

The results of the one-way ANOVA indicate that the means between the three phases are statistically different, and the null hypothesis can be rejected. Note that the significance value is in scientific notation (2.65e-06). Next to this value are three asterisks, indicating the p value is less than .001. Note that the difference in the phase means are reported in alphabetical order. The mean drops 13 points from the baseline to the intervention, is reduced another 6.8 points on intensifying the intervention, and drops a total of 19.8 points overall.

Because the one-way ANOVA is a single test, it is not known which mean differences are accounting for the significant finding (B_1-A, B_2-A, and/or B_2-B_1). The Tukey multiple-comparison test accomplishes this. The Tukey test compensates for the increase in the chance of a Type I error by adjusting the p values. The results

```
> ABanova(depress, pdepress)
                    Df Sum Sq Mean Sq F value   Pr(>F)
as.factor(phaseX)  2 1307.0   653.5   24.35 2.65e-06 ***
Residuals         22  590.4    26.8
---
Signif. codes:  0 '***' 0.001 '**' 0.01 '*' 0.05 '.' 0.1 ' ' 1
2 observations deleted due to missingness
   A   B1   B2
37.8 24.8 18.0
  Tukey multiple comparisons of means
     95% family-wise confidence level

Fit: aov(formula = behavior ~ as.factor(phaseX))

$`as.factor(phaseX)`
       diff       lwr        upr      p adj
B1-A  -13.0 -20.12776  -5.872241 0.0004126
B2-A  -19.8 -26.92776 -12.672241 0.0000015
B2-B1  -6.8 -12.61979  -0.980209 0.0201179
```

Figure 5.20 Console output results of one-way ANOVA comparing depression across phases

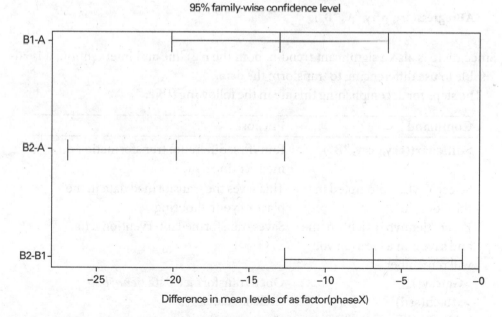

Figure 5.21 Visual comparison of mean depression across phases

indicate all pairs are significantly different from one another. By examining the means, we can conclude that Gloria's depression improved during the intervention phase and improved further once the intensity of the intervention increased.

Tests of Type I Error When Autocorrelation Is High

As previously stated, prior to conducting tests of Type I error, we recommend testing for autocorrelation in both phases. As an example of a method for dealing with the issue of autocorrelation when comparing phases, we look more closely at Jenny's crying behavior.

Begin by using the **ABrf2()** function to assess for autocorrelation in both the baseline and intervention phases by entering the following into the Console after opening and attaching Jenny's file:

>ABrf2(cry,pcry, "A")
>ABrf2(cry,pcry, "B")

The resulting r_{f2} values are 0.302 (*sig* = .425) and 1.019 (*sig* = .00), respectively. The results indicate that the data for the intervention are highly autocorrelated, so it would be advisable to see if transforming the data for this phase reduces or eliminates the issue. Before deciding how to deal with this, check whether the data trend by entering the following in the Console:

>ABregress(cry, pcry, "A", "B")

Since there is also a significant trend in both the baseline and intervention, it is advisable to use differencing to transform the data.

The steps for accomplishing this are in the following table:

Command	Purpose
>diffchart(cry,pcry, "B")	Runs first difference transformation on intervention data.
Select Y when prompted in the Console.	This saves the transformed data in the place of your choosing.
Enter "tjennyb" as a file name and save it in a location you will remember.	Saves transformed intervention data.
>Getcsv()	Open transformed data *tjennyb*.
>attach(ssd)	Attaches data.
>ABrf2(diff, phase, "B")	Runs autocorrelation on transformed intervention data.

NOTE: When you transform data, the behavior variable is renamed *diff*, and the phase variable is renamed *phase*.

The output from the **ABrf2()** on the transformed data shows that the $r_{f2} = 0.209$ and *sig* = .434. It appears as if transforming the data was helpful in decreasing the autocorrelation. In order to proceed we now need to transform the baseline data using the same method we used for transforming the intervention data and test the transformed data for autocorrelation. The steps for accomplishing this are shown in the next table:

Command	Purpose
>Getcsv()	Open *jennyab.csv* data to transform the baseline.
>attach(ssd)	Attaches data.
>diffchart(cry,pcry, "A")	Runs first difference transformation on baseline data.
Select Y when prompted in the Console.	This saves the transformed data in the place of your choosing.
Enter "tjennya" as a file name and save it in a location you will remember.	Saves transformed baseline data.
>Getcsv()	Open transformed data *tjennya*.
>attach(ssd)	Attaches data.
>ABrf2(diff, phase, "A")	Runs autocorrelation on transformed data.

The transformation of the baseline data shows that the $r_{f2} = -0.617$; however, that autocorrelation that exists is not significant (*sig* of $r_{f2} = .109$).

We now merge the two transformed datasets and use this to examine whether there are significant differences between baseline and intervention. To combine the two datasets, enter the commands in the following table:

Command	Purpose
>Append()	Open transformed baseline data first: *tjennya*. Then open the transformed intervention file *tjennyb*. Now save the combined file as *tjennyab*.
>Getcsv()	Open the *tjennyab.csv* file.
>attach(ssd)	Attaches data.

NOTE: When you use the **Append()** function, the behavior variables remain named *diff*, and the phase variables remain named *phase*.

Because there seems to be no significant problem of autocorrelation in either phase with the transformed data, we can proceed with the *t* test after the baseline and intervention phase data are merged since we have already determined that there is not a trend in either the baseline or intervention phases.

In order to conduct the *t* test on the transformed dataset, enter the following command in the Console:

>**ABttest(diff, phase, "A", "B")**

The results of the *t* test on the transformed data are presented in Figures 5.22 and 5.23 (see p. 119).

```
> ABttest(diff, phase, "A", "B")

        Two Sample t-test

data:  A and B
t = 1.1054, df = 31, p-value = 0.2775
alternative hypothesis: true difference in means is not equal to 0
95 percent confidence interval:
 -0.6941684  2.3370255
sample estimates:
 mean of x  mean of y
 0.5833333 -0.2380952


        F test to compare two variances

data:  A and B
F = 10.18, num df = 11, denom df = 20, p-value = 1.157e-05
alternative hypothesis: true ratio of variances is not equal to 1
95 percent confidence interval:
  3.741567 32.843084
sample estimates:
ratio of variances
          10.18029


        Welch Two Sample t-test

data:  A and B
t = 0.87197, df = 12.248, p-value = 0.4
alternative hypothesis: true difference in means is not equal to 0
95 percent confidence interval:
 -1.226499  2.869356
sample estimates:
 mean of x  mean of y
 0.5833333 -0.2380952
```

Figure 5.22 Console output results of t-test comparing differenced crying across phases

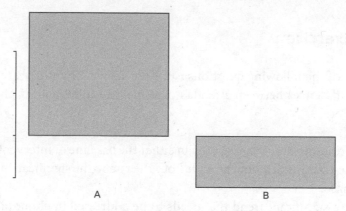

A   B

Figure 5.23 Visual comparison of mean differenced crying across phases

When looking at the test to examine the equality of variances, we see that they are significantly different ($p < .05$), so we look at the t-test results for unequal variances. When we look at these results, we note $p = .4$. As a result, we cannot reject the null hypothesis that the difference in the means between the baseline and intervention is greater than zero. Therefore, we would have to conclude that, statistically, the intervention did not have an impact on Jenny's crying behavior, although it appears as if change is going in the right direction.

Conclusion

In this chapter, we covered the use of a number of hyothesis tests to test for Type I error between phases. SPC charts have the advantage of testing for Type I error while producing graphs that are easily understood. We also covered other statistical tests, including proportion/frequency, chi-square, the CDC, the t test, and ANOVA. While these tests exist, it is important to consider the conditions under which it is appropriate to use each of these, and we recommend using the decision trees in Appendix C to help you select the test of Type I error that is most suitable for your data.

In single-subject research, we recommend using hypothesis tests in conjunction with effect sizes to holistically assess your work with clients. While statistical tests alone are sometimes used to examine change in all types of evaluations, using a combination of both will likely lead to better practical decision-making.

Finally, a method for transforming data and merging datasets was discussed. Once new datasets are created using the **Append()** function, they can be tested for Type I error using the techniques discussed in the chapter.

Chapter Exercise

Answer each of the following questions to determine whether there are statistically significant differences between Brenda's baseline and intervention phase oppositional behavior:

1. Are there issues of autocorrelation in either the baseline or intervention phases? To make your justification, provide both the r_{f2} and the significance of the r_{f2} for each phase.
2. Is there a significant trend that needs to be addressed in either phase? Justify your answer.
3. Conduct an appropriate hypothesis test with justification for why you are using a particular test. Is there a significant difference in Brenda's behavior after the introduction of the intervention?
4. What is the effect of the intervention? Choose an appropriate effect size with justification on why you are utilizing that type of descriptive statistic. Interpret your findings.
5. Assume you are working with Brenda. Based on your findings, what practice decisions would you make? Justify your rationale with the results of your evaluation.

6
Analyzing Group Data

Introduction

Macgowan (2008) pointed out the importance of measuring performance in evidence-based group work (Macgowan, 2008, 2012). A set of functions exists in *SSD for R* for analyzing measures of group behaviors.

This chapter discusses how to enter data on a group target behavior into Excel or any other program that can export data to the *.csv* format. It also explains how to use the *SSD for R* functions to analyze group outcomes.

Entering Group Data

Except for a few extra steps, entering data about groups is very similar to the method discussed in Chapter 1. The steps to do this are presented next:

1. Open Excel or whatever program you are using to create your *.csv* file.
2. On the first row (labeled "1"), enter the names of your variables across the columns, beginning with column A.

You will need to create several variables in Excel: (a) a group behavior variable that corresponds to the group behavior that you are measuring; (b) a phase variable that corresponds to the group behavior variable (the phase variable will indicate in which phase, baseline or intervention, the measurement occurred); (c) a time unit variable in which the behavior occurred (this could represent the group session in which the measurement was taken, e.g., "pweek"); and (d) a group member variable. This variable will indicate which member of the group is being measured for each occurrence.

In order to do this systematically, we recommend giving the group behavior variable a meaningful name; the phase variable name should be similar to the behavior variable name. For example, if you are measuring the degree of group mutual aid, your behavior variable could be named "muaid" and your phase variable could be named "pmuaid." In the example presented in this chapter, we used pweek as our time unit variable to denote that the groups were conducted weekly. Finally, we used "member" to represent a group participant. Starting in row 2, begin entering your data for each occurrence the behavior is measured.

IMPORTANT NOTE: When phases change, you will need to enter "NA" into the row between a change in phases for each variable.

SSD for R. Charles Auerbach and Wendy Zeitlin, Oxford University Press. © Oxford University Press 2022.
DOI: 10.1093/oso/9780197582756.003.0007

In Figure 6.1, note that we have entered baseline and intervention data for a mutual aid group. Note that there are eight group members (labeled as member 1 through 8 in column D) within each week (labeled pweek in column B) displayed in Figure 6.1. The data are entered in order of the time unit as this is the order in which data for this group were collected.

You can also track group members' individual behaviors in the same spreadsheet where you are tracking your group behavior variable. These behaviors are typically not the same as the behavior that is measured for the entire group, and specific behaviors measured can differ for each group member. Note in Figure 6.1 the eight behavior variables m1, m2, . . . m8 and the eight phase variables pm1, pm2, . . . pm8. The column labeled m1 is a behavior variable for the first group member, and pm1 is a phase variable for the first group member. All eight members have individual behaviors recorded. The data were entered using the same steps discussed in Chapter 1.

3. Once your data are entered into Excel, you will need to save it as a "CSV (Comma delimited)" file or "CSV (Comma Separated Values)" in your *ssddata* directory. To do this, choose SAVE AS and choose a name for your file. Do

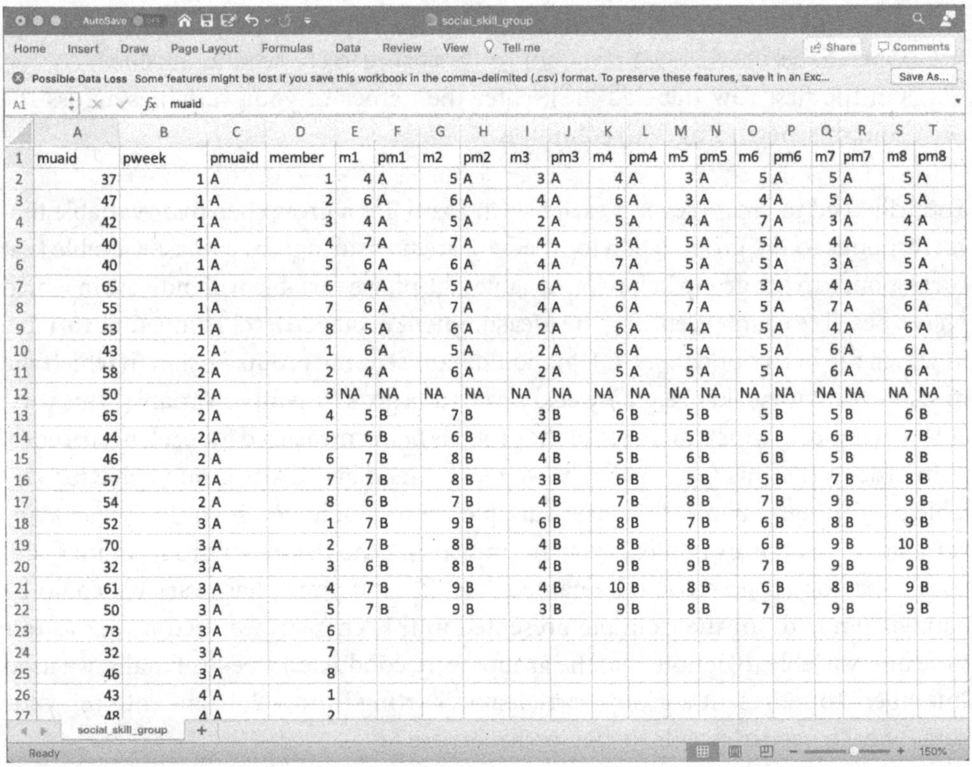

	A	B	C	D	E	F	G	H	I	J	K	L	M	N	O	P	Q	R	S	T
1	muaid	pweek	pmuaid	member	m1	pm1	m2	pm2	m3	pm3	m4	pm4	m5	pm5	m6	pm6	m7	pm7	m8	pm8
2	37	1	A	1	4	A	5	A	3	A	4	A	3	A	5	A	5	A	5	A
3	47	1	A	2	6	A	6	A	4	A	6	A	3	A	4	A	5	A	5	A
4	42	1	A	3	4	A	5	A	2	A	5	A	4	A	7	A	3	A	4	A
5	40	1	A	4	7	A	7	A	4	A	3	A	5	A	5	A	4	A	5	A
6	40	1	A	5	6	A	6	A	4	A	7	A	5	A	5	A	3	A	5	A
7	65	1	A	6	4	A	5	A	6	A	5	A	4	A	3	A	4	A	3	A
8	55	1	A	7	6	A	7	A	4	A	6	A	7	A	5	A	7	A	6	A
9	53	1	A	8	5	A	6	A	2	A	6	A	4	A	5	A	4	A	6	A
10	43	2	A	1	6	A	5	A	2	A	6	A	5	A	5	A	6	A	6	A
11	58	2	A	2	4	A	6	A	2	A	5	A	5	A	5	A	6	A	6	A
12	50	2	A	3	NA	NA	NA	NA	NA	NA	NA	NA	NA	NA	NA	NA	NA	NA	NA	NA
13	65	2	A	4	5	B	7	B	3	B	6	B	5	B	5	B	5	B	6	B
14	44	2	A	5	6	B	6	B	4	B	7	B	5	B	5	B	6	B	7	B
15	46	2	A	6	7	B	8	B	4	B	5	B	6	B	6	B	5	B	8	B
16	57	2	A	7	7	B	8	B	3	B	6	B	5	B	5	B	7	B	8	B
17	54	2	A	8	6	B	7	B	4	B	7	B	8	B	7	B	9	B	9	B
18	52	3	A	1	7	B	9	B	6	B	8	B	7	B	6	B	8	B	9	B
19	70	3	A	2	7	B	8	B	4	B	8	B	8	B	6	B	9	B	10	B
20	32	3	A	3	6	B	8	B	4	B	9	B	9	B	7	B	9	B	9	B
21	61	3	A	4	7	B	9	B	4	B	10	B	8	B	6	B	8	B	9	B
22	50	3	A	5	7	B	9	B	3	B	9	B	8	B	7	B	9	B	9	B
23	73	3	A	6																
24	32	3	A	7																
25	46	3	A	8																
26	43	4	A	1																
27	48	4	A	2																

Figure 6.1 Example for entering group data

NOT click SAVE, but instead select one of the CSV options from the drop-down menu for SAVE AS TYPE or FORMAT. After you finish this, you should click SAVE and close Excel. You may receive several warnings, but you can accept all of these by selecting CONTINUE.

4. Once you enter your data into Excel, you can import it into *SSD for R*. First, remember to load the *SSD for R* package by typing **require(SSDforR)** in the Console and pressing the <RETURN> key. You can now begin your analysis using the **Getcsv()** function. Once the file is open, type **attach(ssd)**.

Now type **listnames()** to obtain a list of variables in the file.

Analyzing Group Data

Many of the same issues discussed in previous chapters with regard to understanding the baseline, using descriptive statistics, and comparing phases are also relevant when analyzing group data.

An example file *social skills group.csv* is used throughout this chapter to illustrate how to analyze group data. This dataset contains a group measure of the degree of mutual aid, with higher scores indicating a higher degree of providing/using mutual aid within the group. An individual behavior, the degree of self-esteem experienced by the individuals in the group, is also measured using a self-anchoring scale ranging from 1 (least) to 10 (highest).

To begin, use **Getcsv()** to open the file. Once the file is open, type **attach(ssd)** <ENTER> followed by **names(ssd)** <ENTER>. The list of variables in the order in which they were entered into Excel is shown in Figure 6.2 (see p. 124) and will be displayed in the Console.

The variable muaid is the group behavior variable, pweek is the time unit variable, pmuaid is the phase variable denoting whether the measurement was made during the baseline or intervention, and member is a number representing the group member being measured. The columns labeled m1 through m8 are the self-esteem measures for each group member. The phase variable associated with m1 is "pm1," m2 is associated with "pm2," and so on. As previously stated, these represent a target behavior (i.e., self-esteem) and phase for each of the individual group members. There are 10 weeks of baseline data and 10 weeks of intervention data.

The first step in analyzing your data is to describe each of the phases. The typical conventions of single-subject research that use visual analysis and descriptive statistics (e.g., mean, median, quantiles) are used to accomplish this; however, because of the nature of group data, some of the functions designed for individual measures do not adequately describe the data, and functions especially designed for groups are used in their place.

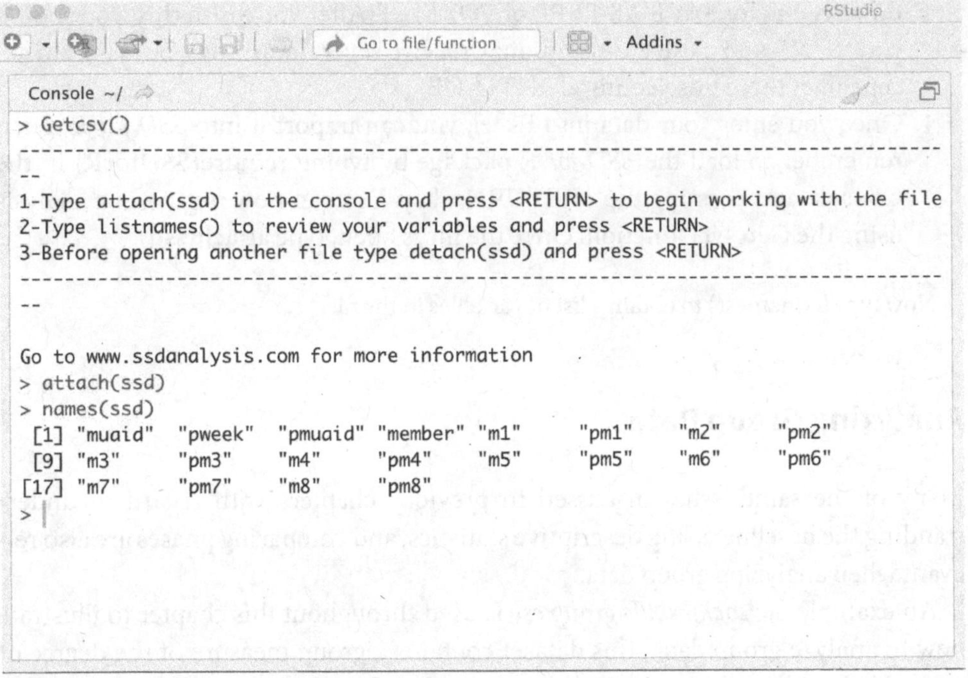

```
● ● ●                                                              RStudio
⊙ ▾ | ⚙ | ⬅ ▾ | ▣ ▣ | ⬇ |  ➤ Go to file/function    | ▦ ▾ Addins ▾

 Console ~/ ⬀                                                        ⬚
> Getcsv()
----------------------------------------------------------------------
--
1-Type attach(ssd) in the console and press <RETURN> to begin working with the file
2-Type listnames() to review your variables and press <RETURN>
3-Before opening another file type detach(ssd) and press <RETURN>
----------------------------------------------------------------------
--

Go to www.ssdanalysis.com for more information
> attach(ssd)
> names(ssd)
 [1] "muaid"   "pweek"   "pmuaid"  "member"  "m1"    "pm1"    "m2"    "pm2"
 [9] "m3"      "pm3"     "m4"      "pm4"     "m5"    "pm5"    "m6"    "pm6"
[17] "m7"      "pm7"     "m8"      "pm8"
> |
```

Figure 6.2 Variable list for social skills group

Descriptive Statistics

It is appropriate to describe the behavior by a time unit variable that, in our example, is weeks. Let's begin by looking at the group behavior by week more closely. A box plot can be created as follows on muaid by pweek by using the following function:

>ABdescrip(muaid, pweek)

To make this a bit more descriptive, a line based on the median for the baseline can be added using the following function:

>Gmedian(muaid,pmuaid,"A")

The **ABlineD(muaid)** can then be used to place a vertical dashed line between phases at the *x*-ordinate of 10.5 because the baseline phase changes to the intervention after the 10th occurrence. The **ABtext()** function can then be used to label the phases. Figure 6.3 (see p. 125) displays this annotated box plot.

The values for the mean, median, standard deviation, quantiles, and other descriptive statistics by week for mutual aid appear in the Console and in Figure 6.4 (see p. 125).

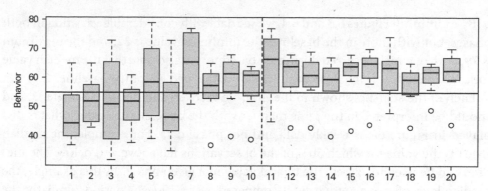

Figure 6.3 Boxplot of mutual aid group with baseline median displayed

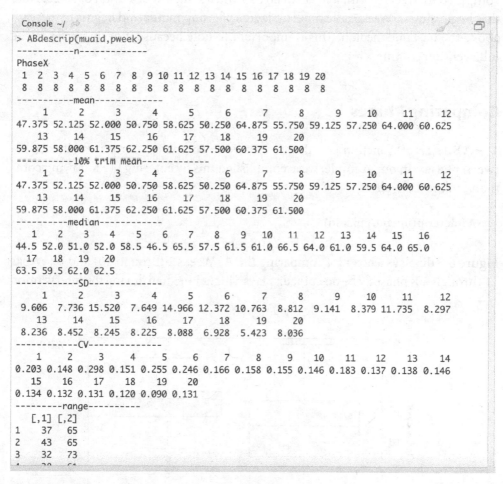

```
Console ~/

> ABdescrip(muaid,pweek)
-----------n-------------
PhaseX
1  2  3  4  5  6  7  8  9 10 11 12 13 14 15 16 17 18 19 20
8  8  8  8  8  8  8  8  8  8  8  8  8  8  8  8  8  8  8  8
-----------mean-------------
      1      2      3      4      5      6      7      8      9     10     11     12
47.375 52.125 52.000 50.750 58.625 50.250 64.875 55.750 59.125 57.250 64.000 60.625
     13     14     15     16     17     18     19     20
59.875 58.000 61.375 62.250 61.625 57.500 60.375 61.500
-----------10% trim mean-------------
      1      2      3      4      5      6      7      8      9     10     11     12
47.375 52.125 52.000 50.750 58.625 50.250 64.875 55.750 59.125 57.250 64.000 60.625
     13     14     15     16     17     18     19     20
59.875 58.000 61.375 62.250 61.625 57.500 60.375 61.500
-----------median-------------
   1    2    3    4    5    6    7    8    9   10   11   12   13   14   15   16
44.5 52.0 51.0 52.0 58.5 46.5 65.5 57.5 61.5 61.0 66.5 64.0 61.0 59.5 64.0 65.0
  17   18   19   20
63.5 59.5 62.0 62.5
------------SD--------------
     1      2      3      4      5      6      7      8      9     10     11     12
 9.606  7.736 15.520  7.649 14.966 12.372 10.763  8.812  9.141  8.379 11.735  8.297
    13     14     15     16     17     18     19     20
 8.236  8.452  8.245  8.225  8.088  6.928  5.423  8.036
------------CV--------------
    1     2     3     4     5     6     7     8     9    10    11    12    13    14
0.203 0.148 0.298 0.151 0.255 0.246 0.166 0.158 0.155 0.146 0.183 0.137 0.138 0.146
   15    16    17    18    19    20
0.134 0.132 0.131 0.120 0.090 0.131
---------range----------
  [,1] [,2]
1   37   65
2   43   65
3   32   73
```

Figure 6.4 Descriptive statistics for mutual aid group

By examining Figures 6.3 and 6.4, we see data with considerable variation in both phases, but with more in the baseline. The minimum value is 32, and the maximum is 79, both of which occurred during the baseline. We also note that the median value for each week during the intervention is above the median for the baseline.

Each of the statistics shown in the Console is covered in detail in Chapter 3, and should be interpreted in the same manner with the understanding that what is displayed in Figure 6.4 is weekly data and not phase data. For example, the median (Md) is the value for which 50% of the observations fall above and below. The median can be used to express the typical value in a given phase. In this example, the median level of group mutual aid is compared over a 10-week period. Similarly, the other measures of central tendency and variation are calculated within each week. They all tend to reflect a great deal of variation between weeks. For example, for Week 3 the mean level of mutual aid is 52.00 with a standard deviation of 15.520 compared to Week 19 with a mean of 60.375 and a standard deviation of 5.423. Note that beginning in Week 7, the median degree of group mutual aid begins to increase above the baseline median. This is important to note because it occurs prior to the intervention commencing.

Comparing Phases

The **ABdescrip()** function can also be utilized to compare descriptive statistics between phases. In our example, we accomplished this by entering the following command in the Console:

>**ABdescrip(muaid, pmuaid)**

Figure 6.5 displays a box plot comparing the A (Weeks 1 through 10) and B (Weeks 11 through 20) phases. The box plot displays a higher median during the intervention

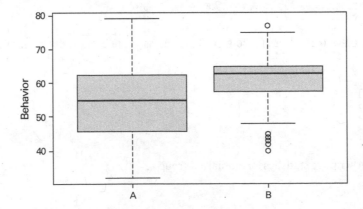

Figure 6.5 Boxplot comparing mutual aid between phases

phase. However, consider the results of Figure 6.3, which shows that the degree of mutual aid had begun to increase during Week 7, prior to the intervention being introduced during Week 11. This example shows the usefulness of displaying group data by week, which provides greater detail than overall phase data, and its use should always be considered.

Figure 6.6 shows the output for the phase data displayed on the Console. From this, we observe that there was a mean increase in mutual aid from baseline to intervention, and, as displayed by the standard deviations, variation decreased during the intervention.

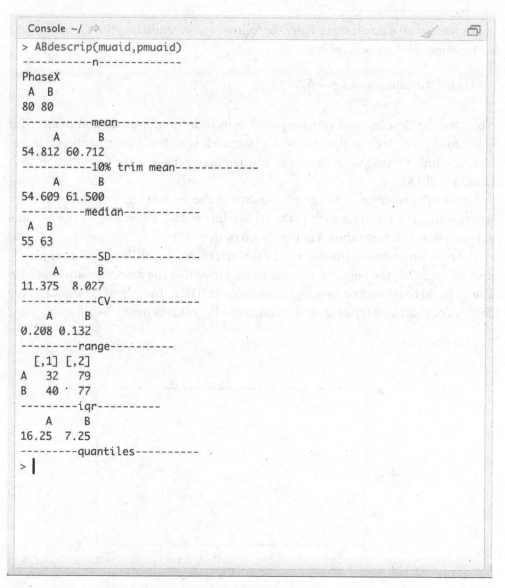

Figure 6.6   Descriptive statistics for mutual aid by phase

Autocorrelation and Trend

As discussed in Chapters 3 and 5, statistical tests rely on various assumptions about the nature of the data being analyzed. Regardless of whether individual or group data are measured, when these assumptions are not met, incorrect conclusions about calculated values can be made, such as deciding that observed differences between phases are significant when they, in fact, are not (i.e., Type I error). On the other hand, we may erroneously fail to detect a difference between phases when those differences do exist (i.e., Type II error). As with individual data, it is important to consider the presence of autocorrelation or trends in any phase before deciding on how to compare phases.

For our group data example, enter the following command into the Console to test for baseline lag-1 autocorrelation:

>GABrf2(muaid,pmuaid,pweek,"A")

Note that the time interval variable pweek is included in this group command. This is because there are multiple measures per week (i.e., the eight members); therefore, the unit of analysis is the mean of members per week (in consultation with Huitema, 2013).

Figure 6.7 presented in the graph window to the bottom right displays a red dot representing the mean for each of the 10 baseline weeks. The results of the baseline autocorrelation test are shown in Figure 6.8 (see p. 129).

The r_{f2} is a measure of the degree of autocorrelation. Regardless of direction, positive or negative, the higher the r_{f2} the more influential the autocorrelation will be. The r_{f2} in this case is close to zero, with a value of 0.092. The "sig of r_{f2}" indicates the chances of making a Type I error with regard to the value of the r_{f2}. Because this value

A

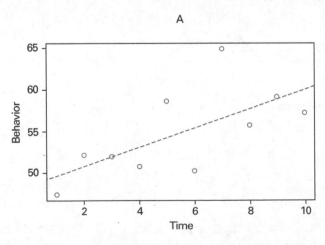

Figure 6.7 Graphical output from **GABrf2()** function for mutual aid baseline example

```
Console ~/

> GABrf2(muaid,pmuaid,pweek,"A")
            1
 "tf2=" "0.197"
            1
 "rf2=" "0.092"
                    1
"sig of rf2="       "0.846"
----------regression------------

Call:
lm(formula = A ~ x1)

Residuals:
    Min     1Q Median     3Q    Max
-5.1402 -2.3068 -0.9375  1.0843  8.3295

Coefficients:
            Estimate Std. Error t value Pr(>|t|)
(Intercept)  48.4583     2.8461  17.026 1.44e-07 ***
x1            1.1553     0.4587   2.519   0.0359 *
---
Signif. codes:  0 '***' 0.001 '**' 0.01 '*' 0.05 '.' 0.1 ' ' 1

Residual standard error: 4.166 on 8 degrees of freedom
Multiple R-squared:  0.4423,    Adjusted R-squared:  0.3726
F-statistic: 6.344 on 1 and 8 DF,  p-value: 0.03588

>
```

Figure 6.8 Test for autocorrelation and trend for mutual aid in baseline

is above 0.05 in our example (sig of r_{f2} = .846), it is not statistically significant. As a result, we can conclude these data are not autocorrelated.

Figure 6.7 displays a moderate trend in the data, with the blue regression line displaying the slope; however, note that the individual data points are not clustered tightly around the regression line. In addition to the graph, additional output is produced in the Console under "regression," as displayed in Figure 6.8. These values show that the degree of change can be quantified by the estimate for x1 of 1.1553, which is the slope of the regression line. This can be interpreted as follows: For every one-unit increase in time (in this case, a week), there is a 1.1553 increase in

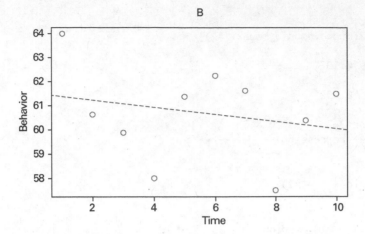

Figure 6.9 Graphical output from **GABrf2()** function for mutual aid intervention example

group mutual aid. The column labeled "t value" is the calculated value for the statistical significance of the slope and constant. The last column, labeled "Pr(>|t|)" is the probability of making a Type I error (i.e., concluding that the coefficient for the slope is greater than 0). Because the probability of the coefficient is less than the commonly accepted threshold of .05, the slope is considered statistically significant. This also demonstrates that even though a trend exists in the data, the data are not autocorrelated, as explained previously.

Similarly, the following command is needed to test for baseline lag-1 autocorrelation for the intervention phase, the results of which are displayed in Figures 6.9 and 6.10 (see p. 131):

>GABrf2(muaid,pmuaid,pweek,"B")

Figure 6.10 (see p. 131) presents the statistical results of the analysis. The r_{f2} for the intervention phase is 0.291. Because the sig of r_{f2} is above 0.05 in our example (sig = .537), it is not statistically significant. As a result, we can conclude the intervention data are not autocorrelated. Figure 6.9 displays a lack of a trend in the data, and the coefficient of −0.1477 [Pr(>|*t*|) > 0.05] indicates little change over time. Despite an insignificant *p* value, however, we may still want to consider the fit of the data around the regression line because, again, significance may be hard to achieve, particularly with small sample sizes.

If the data are autocorrelated or you are assuming autocorrelation, you could attempt to transform the phase data in order to reduce the effect of autocorrelation. If the data have a high degree of variation, using a moving average could smooth them and reduce the impact of autocorrelation. The moving average is just the plot of the mean of two adjacent observations: ([point 1 + point 2]/2). Alternatively, if the data are autocorrelated and have a trend, attempting to transform the data using differencing is recommended. For more information on using the moving average or differencing functions in *SSD for R*, refer to Chapter 3.

```
Console ~/

> GABrf2(muaid,pmuaid,pweek,"B")
             11
 "tf2=" "0.632"
             11
 "rf2=" "0.291"
                      11
"sig of rf2="         "0.537"
----------regression------------

Call:
lm(formula = A ~ x1)

Residuals:
    Min      1Q  Median      3Q     Max
-2.9341 -1.0562  0.3841  1.3727  2.6227

Coefficients:
            Estimate Std. Error t value Pr(>|t|)
(Intercept)  61.5250     1.3612  45.199 6.34e-11 ***
x1           -0.1477     0.2194  -0.673     0.52
---
Signif. codes:  0 '***' 0.001 '**' 0.01 '*' 0.05 '.' 0.1 ' ' 1

Residual standard error: 1.993 on 8 degrees of freedom
Multiple R-squared:  0.05364,   Adjusted R-squared:  -0.06465
F-statistic: 0.4535 on 1 and 8 DF,  p-value: 0.5197

>
```

Figure 6.10   Test for autocorrelation and trend for mutual aid in intervention

Using Effect Size to Describe Group Change

Thus far we have described visual techniques combined with measures of central tendency and variation to describe change in behavior across phases. Effect size quantifies the amount of change between phases. Measuring effect size has become a common method for assessing the degree of change between phases as it gives a sense of practical, or clinical, significance (Ferguson, 2009; J. G. Orme, 1991). When effect size is used to describe data, it has advantages over tests of statistical signifi-cance in that the focus is on the magnitude of change between phases rather than whether differences are statistically significant (Kromrey & Foster-Johnson, 1996). An additional advantage is that effect size is also relatively easy to interpret. *SSD for R* calculates two forms of effect size that are appropriate for group data: ES and d-index.

Entering the following command in the Console will produce the output displayed in Figure 6.11 (see p. 133).

>Effectsize(muaid,pmuaid,"A","B")

Notice the request to choose one of the following: (s)ave, (a)ppend, or (n)either results? (s/a or n); select **n**. The other two choices are used to store the effect size results to a file for use in meta-analysis.

All effect sizes ES (0.51867), d-index (0. 59932), and Hedge's g (0.59647) indicate a small degree of change, as noted in the interpretation of effect sizes toward the top of Figure 6.11. The percentages of change for the ES and d-index are 19.8% and −22.55%, respectively.

Tests of Statistical Significance

SSD for R contains two tests of statistical significance that can be applied to group data. There is a function that is designed specifically to do a *t* test using group data, which differs from the *t* test described Chapter 5. The proportion/frequency test, as described in Chapter 5, can also be used with group data. Additionally, three statistical process control (SPC) charts, the X-bar R chart ($\bar{X}$-R chart) and two R charts are also appropriate for the analysis of group data.

t-test

The *t* test compares the differences in the means between phases. The null hypothesis is that the difference in means between the phases is zero (i.e., there is no change). If the *t* test indicates statistically significant differences between the phases, that is, $p \leq .05$, then we accept the alternative that the mean differences are greater than zero. The *t* test, however, is nondirectional, meaning that significant *p* values could indicate either improvement or deterioration in the observed behavior. If higher values are desired, the difference should be positive, and if lower values are desired, the difference should be negative.

As mentioned in Chapter 5, there are some caveats for the use of the *t* test. It should not be used if there is a trend in the data or if either phase has problematic autocorrelation. If the data are autocorrelated, a transformation of the data can be used to try to remove it. If the problem of autocorrelation is resolved, the transformed data can then be analyzed using the *t* test.

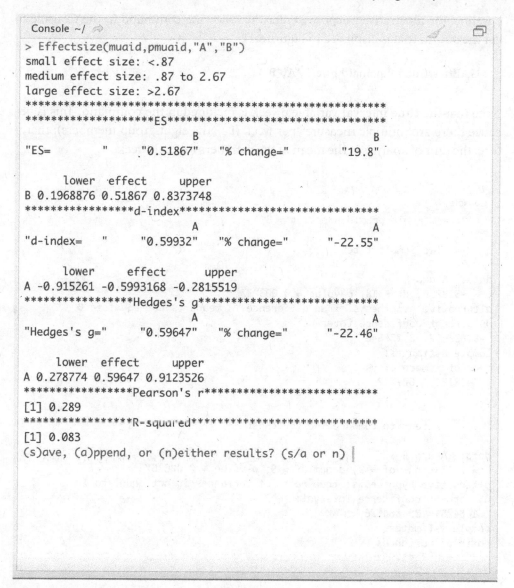

```
Console ~/
> Effectsize(muaid,pmuaid,"A","B")
small effect size: <.87
medium effect size: .87 to 2.67
large effect size: >2.67
***********************************************************
*********************ES************************************
                           B                          B
"ES=          "      "0.51867"   "% change="        "19.8"

       lower  effect     upper
B 0.1968876 0.51867 0.8373748
****************d-index****************************
                           A                        A
"d-index=  "        "0.59932"   "% change="    "-22.55"

       lower      effect       upper
A -0.915261 -0.5993168 -0.2815519
****************Hedges's g*************************
                          A                      A
"Hedges's g="      "0.59647"   "% change="    "-22.46"

       lower   effect      upper
A 0.278774 0.59647 0.9123526
****************Pearson's r***********************
[1] 0.289
****************R-squared*************************
[1] 0.083
(s)ave, (a)ppend, or (n)either results? (s/a or n) |
```

Figure 6.11 Effect size calculation for mutual aid group

The *t* test also assumes equal variation between the phases. If the variances are unequal, then a *t* test for unequal variances should be used. The group *t*-test function in *SSD for R* provides a test for the differences in variances between phases, which can be interpreted to determine which form of the *t* test should be used. Additionally, output for the group *t*-test function includes results for both equal and unequal variances, and the appropriate test results can then be analyzed.

Looking at our example of the mutual aid group, we recall that the baseline data are not autocorrelated but do have a moderate trend. Although there is a trend in the baseline, for illustrative purposes, enter the following command in the Console to get the output that is displayed in Figure 6.12:

>GABttest(muaid,pmuaid,pweek,"A","B")

Note that the time interval variable pweek is included in the command. This is because there are multiple measures per week (i.e., the eight group members); therefore, the unit of analysis is the mean of the members for each week.

```
Console ~/
> GABttest(muaid,pmuaid,pweek,"A","B")

        Two Sample t-test

data:  A and B
t = -3.3299, df = 18, p-value = 0.003727
alternative hypothesis: true difference in means is not equal to 0
95 percent confidence interval:
 -9.622452 -2.177548
sample estimates:
mean of x mean of y
  54.8125    60.7125


        F test to compare two variances

data:  A and B
F = 7.418, num df = 9, denom df = 9, p-value = 0.006367
alternative hypothesis: true ratio of variances is not equal to 1
95 percent confidence interval:
  1.842514 29.864626
sample estimates:
ratio of variances
         7.417951


        Welch Two Sample t-test

data:  A and B
t = -3.3299, df = 11.383, p-value = 0.006419
alternative hypothesis: true difference in means is not equal to 0
95 percent confidence interval:
 -9.783787 -2.016213
sample estimates:
mean of x mean of y
  54.8125    60.7125

>
```

Figure 6.12 *t*-test comparing baseline and intervention for mutual aid group

There are three tests displayed in the Console panel to the left of the graphical output: a *t* test for equal variances at the top of the Console, an F test to compare the variances in the middle, and a *t* test for unequal variances at the bottom. To begin, look at the *p* value for the F test to compare the variances. Because the *p* value for equality variances in our example is less than .05 (.006367), we reject the null hypothesis that the variances are equal. Therefore, we use the *t*-test results for unequal variances to determine whether there are statistically significant differences between the baseline and intervention phases.

Looking at the results for unequal variances, located at the bottom of the Console, we reject the null hypothesis because the *p* value is less than .05 (.0006419). Therefore, we accept the alternative that the mean difference between the phases is greater than zero. We also observe that the mean went up from 54.8125 in the baseline to 60.7125 in the intervention.

Proportion/Frequency

If you are using task completion data or any binary outcome, the **ABbinomial()** function can be utilized to test for statistical significance between phases. This command is discussed at length in Chapter 5. No adjustments are needed to this function for group data.

Using an X̄-R Chart to Visualize Group Change

The X̄-R chart can be used with group data because there are multiple observations per sample (J. G. Orme & Cox, 2001). In the case of our mutual aid group example, the group data were collected on a weekly basis, which resulted in 20 samples (i.e., 10 weeks of baseline plus 10 weeks of intervention), with eight observations (i.e., group members) per sample. Using the mean of these samples increases the accuracy of our measure, which are the individual data points displayed in the X̄-R chart (Mohammed & Worthington, 2012; Orme & Cox, 2001).

To create the X̄-R chart displayed in Figure 6.13 (see p. 136), enter the *SSD for R* commands presented in the following table in the Console:

Command	Purpose
>XRchart(muaid,pweek,2,"week","Level of >Mutual Aid","Group Mutual Aid")	Creates graph using two standard deviations
>ABlines(muaid)	Adds line between phases
>ABtext("57.763")	Enters mean text
>ABtext("64.605")	Enters Uband (upper band) text
>ABtext("50.920")	Enters Lband (lower band) text
>SPClegend()	Creates a legend

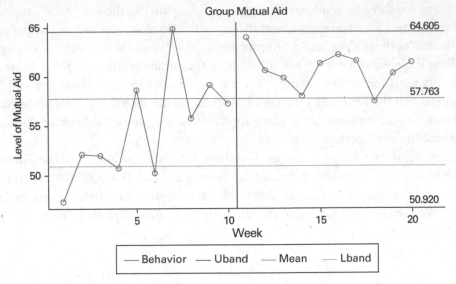

Figure 6.13 X-R chart for mutual aid grou

NOTE: Once you create a legend, you will no longer be able to alter the graph. Therefore, we add this as a last step in creating SPC charts. Be sure that your graph looks exactly as you would like before adding the legend.

In this example, the desired zone would be above the upper band (64.605) since the group's goal is to increase the level of mutual aid. Although 9 of 10 weeks during the intervention are above the mean, none of the intervention observations is in the desired zone.

In terms of practice decisions, the group facilitator would probably want to continue measuring this process because it seems that the intervention is going in the desired direction. To increase confidence, continued tracking would be recommended.

Using R Charts to Assess Variability of Group Data

As discussed in Chapter 4, R charts can be used to detect changes in variation over time. As these SPC charts use samples, they are appropriate for use with group data. In the example of our mutual aid group, it is most appropriate to use the standard deviation form of the R chart, which is illustrated in Figure 6.14 (see p. 137). To create this chart with two standard deviations, the following command was entered into the Console:

>**Rchartsd(muaid, pweek, 2, "week", "standard deviation", "Group Mutual Aid")**

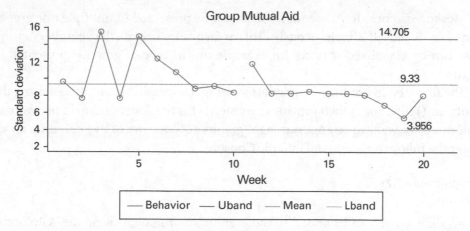

Figure 6.14 R-chart using standard deviations for mutual aid group

The graph was further annotated by noting values for the mean, upper, and lower bands using the **ABtext**() function and adding a legend in the same way that the X̄-R chart was annotated, above, with the **SPClegend**() function.

In looking at the output, we see that the data became more stable in the second half of the baseline and throughout the intervention. Combined with our previous analysis, this could help us conclude that, over time, the group's level of mutual aid was moving in the desired direction, and that, as a whole, the overall level of mutual aid was beginning to become less variable between group members.

Using P Charts to Assess Binary Outcomes in Group Data

If you recall, P charts can be used to assess binary outcomes between phases. P charts work well with group data as they use a grouping variable (e.g., weeks). While our sample dataset does not contain any binary outcomes, this SPC chart can be used with group data in the way that it was described in Chapter 4.

Individual Change

As mentioned, the individual change of group members can also be measured with *SSD for R*. In our hypothetical example, the group members used a self-anchoring scale weekly to rate their self-esteem. The scale ranged from a low of 1 to a high of 10. The measured score for each group member is contained in the columns labeled m1 to m8. The corresponding phase variables for these are contained in columns labeled pm1 to pm8.

Measured scores for these individuals can be presented in multiple line graphs that are displayed simultaneously. This is helpful since all the members' progress can be visualized at once. An example of this type of graph is presented in Figure 6.15.

The first step in creating the graph is to set up the graphic environment using the **plotnum**() function, which provides parameters for the desired number of rows and columns. We want to arrange our eight graphs as four rows of two columns, so we enter the following command into the Console:

>**plotnum(4,2)**

Now, each graph can be added to the graphic environment using the **ABplotm**() commands seen in the next table. Note that this function is different from the **ABplot**() function:

NOTE: Once you create an individual graph, you will able to alter and annotate the graph until the next graph is added. Be sure that your graph looks exactly as you would like it before adding another. If the following error occurs, "figure margins too large," you will have to increase the size of the Plots pane. If this does not solve the issue, reset the Plots pane by issuing the following *R* command in the Console: **dev.off**(). If you do this, be aware that any graphs currently in the pane will be removed.

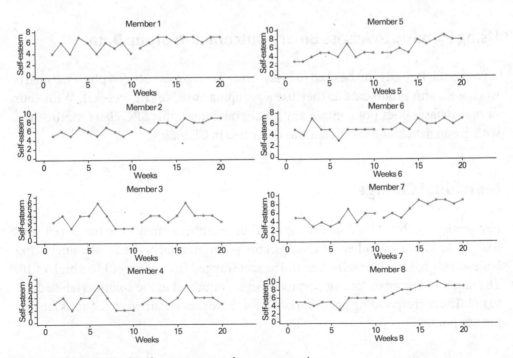

Figure 6.15 Individual self-esteem scores for group members

Command	Purpose
>ABplotm(m1,pm1,"weeks","self esteem","member 1")	Creates graph
>ABlines(m1)	Adds line between phases
>ABplotm(m2,pm2,"weeks","self esteem","member 2")	Creates graph
>ABlines(m2)	Adds line between phases
>ABplotm(m3,pm3,"weeks","self esteem","member 3")	Creates graph
>ABlines(m3)	Adds line between phases
>ABplotm(m4,pm4,"weeks","self esteem","member 4")	Creates graph
>ABlines(m4)	Adds line between phases
>ABplotm(m5,pm5,"weeks","self esteem","member 5")	Creates graph
>ABlines(m5)	Adds line between phases
>ABplotm(m6,pm6,"weeks","self esteem","member 6")	Creates graph
>ABlines(m6)	Adds line between phases
>ABplotm(m7,pm7,"weeks","self esteem","member 7")	Creates graph
>ABlines(m7)	Adds line between phases
>ABplotm(m8,pm8,"weeks","self esteem","member")	Creates graph
>ABlines(m8)	Adds line between phases

These individual data can be analyzed in the same manner we discussed in previous chapters. For example, you can obtain descriptive statistics for each member using the **ABdescrip**() command. If the necessary assumptions are met, you can calculate effect size and tests of statistical significance and can create any of the graphs discussed previously. All the assumptions discussed in the previous chapters need to be considered, such as baseline stability, autocorrelation, trend, and standard deviation.

Conclusion

In this chapter, we discussed how you could use *SSD for R* to assess group functioning. There are functions to compare change over time or between phases. Using the group social interaction example data, you learned the functions necessary to calculate these statistics. These functions generate supporting graphs, box plots, SPC charts, and individual line graphs. *SSD for R* also provides methods for computing group autocorrelation, group effect size, and group *t* tests.

Chapter Exercises

Assignment 6.1—Create a Group Spreadsheet and Import It to *RStudio*

A social worker serving mental health clients is assigned a group of four clients having difficulty adapting to transitional housing because of the quality of their activities of living (ADL) skills, such as cleaning, laundry, maintaining their residences, shopping for food, preparing meals, paying bills, and taking public transportation. The worker decides to start a group and provide an intervention to teach the clients ADL skills. She measures their ADL skills at each session with a scale ranging from 1 (*poor*) to 10 (*excellent*).

For this assignment, you will create a spreadsheet with the group's baseline and intervention data. BE SURE TO STORE THIS SPREADSHEET IN A PLACE YOU CAN ACCESS LATER, as you will use it for future homework assignments. The data you need for this is in the following table:

ADL Score	Week of Service	Phase (A or B)	Member ID
4	1	A	1
4	1	A	2
6	1	A	3
3	1	A	4
2	2	A	1
3	2	A	2
4	2	A	3
2	2	A	4
7	3	A	1
3	3	A	2
2	3	A	3
3	3	A	4
3	4	A	1
4	4	A	2
5	4	A	3
3	4	A	4
2	5	A	1
2	5	A	2
4	5	A	3
6	5	A	4
NA	NA	NA	NA
5	6	B	1

ADL Score	Week of Service	Phase (A or B)	Member ID
7	6	B	2
8	6	B	3
4	6	B	4
6	7	B	1
5	7	B	2
7	7	B	3
6	7	B	4
6	8	B	1
7	8	B	2
6	8	B	3
5	8	B	4
7	9	B	1
6	9	B	2
7	9	B	3
6	9	B	4
6	10	B	1
7	10	B	2
8	10	B	3
6	10	B	4
7	11	B	1
6	11	B	2
5	11	B	3
7	11	B	4
8	12	B	1
6	12	B	2
7	12	B	3
5	12	B	4
6	13	B	1
7	13	B	2
7	13	B	3
7	13	B	4
8	14	B	1
6	14	B	2
7	14	B	3
6	14	B	4
7	15	B	1
7	15	B	2
8	15	B	3
8	15	B	4

142 *SSD for R*

Assignment 6.2—Visual Analysis

1. Create an $\bar{\text{X}}$-R chart, labeling it appropriately. In the main title, include YOUR initials. For example, my main title could read, "Group Memory—CA."
2. Annotate the line graph to put a vertical line between phases.
3. Label the baseline phase "A" and the intervention phase "B."
4. Export the graph to a Word document and add your answer to the following question: Based on the line graph alone, do you think that the group intervention with the ADL group is having the desired effect? Why or why not?

Assignment 6.3—Hypothesis Testing and Effect Sizes

Answer each of the following questions to determine whether there are significant differences between the group's baseline and intervention phase memory behavior:

1. Are there issues of autocorrelation in either the baseline or intervention phases? To make your justification, provide both the r_{f2} and the significance of the r_{f2} for each phase.
2. Is there a significant trend that needs to be addressed in either phase? Justify your answer.
3. Conduct an appropriate hypothesis test with justification for why you are using a particular test. Is there a significant difference in memory after the introduction of the intervention?
4. What is the effect of the intervention? Choose an appropriate effect size with justification for why you are utilizing that type of descriptive statistic. Interpret your findings.
5. Assume you are responsible for this group. Based on your findings, what practice decisions would you make? Justify your rationale with the results of your evaluation.

7

Meta-Analysis in Single-Subject Evaluation Research

Introduction to Meta-Analysis

Throughout this book, we have discussed methods for evaluating the degree to which change can be observed and interpreted with the introduction of intervention(s) to help remediate some identified problem(s). The findings from these evaluations should inform your work with individual client systems, but sometimes, we might want to know more.

Oftentimes our work involves many client systems with similar presenting problems, and we might use similar intervention techniques to help address those problems. We also know that some client systems respond better to intervention than others. So, we might wonder, given clients with a particular type of presenting problem (e.g., depression), how much change could we expect to see if we use a particular intervention (e.g., cognitive behavioral therapy)? And, does that change more generally make meaningful differences in the lives of our clients?

Meta-analytic techniques can be used to aggregate evaluation results across studies. In the case of single-subject research designs, we could combine findings from evaluations with 5, 10, or 20 clients to determine, on average, how effective an intervention is. We can combine effect sizes discussed in Chapter 4 to report the direction and magnitude of change across all those clients (i.e., individual studies). This is a more complex and sophisticated way of understanding differences across studies than reporting those changes qualitatively or simply reporting the individual effect sizes for each study.

Meta-analysis is, in essence, a study of studies. It is most frequently used to understand average effects of interventions across research in which group designs are used, and it is predicated on the assumption that individual studies are independent of one another. Therefore, it is only appropriate to use meta-analytic techniques with different clients. That is, do not conduct a meta-analysis with the same client across different outcomes.

In this chapter, we talk more about why meta-analysis is important to consider in single-subject research, and then we demonstrate how to do this using *SSD for R* functions.

SSD for R. Charles Auerbach and Wendy Zeitlin, Oxford University Press. © Oxford University Press 2022.
DOI: 10.1093/oso/9780197582756.003.0008

How Do I Conduct a Meta-Analysis?

As with other aspects of this text, we do not go into great detail on how to conduct a thorough meta-analysis as many excellent books and resources are available on this in great depth. Some of these resources are listed in Appendix D because we find them particularly relevant to single-subject case designs.

In short, however, meta-analysis is not unlike methods used in other research. For example, you should first start with an answerable research question. Instead of generating data by conducting original research, meta-analysis is dependent on conducting a systematic search for studies that can help answer the research question and often entails a specification of inclusion and exclusion criteria. That is, the researcher will formalize a rationale for why studies should be included or left out of a meta-analysis.

Then, data have to be extracted from studies included in the meta-analysis. This could include either the raw data used to compute effect sizes or the effect sizes themselves along with other variables that could be moderators. From there, data can be analyzed and interpreted based on output from these analyses.

Why Is Meta-Analysis Important in Single-Subject Research?

Single-subject research designs are wonderful ways to consider change across individual client systems, but they are not generally considered particularly rigorous research designs in the scheme of published research. This is because it is impossible to generalize findings of these studies to larger populations because we only have a sample size of one in these designs. Meta-analysis, then, enables researchers to combine these findings across studies to begin to build generalizability to larger groups.

The What Works Clearinghouse Single-Case Design Standards recommend combining a minimum of five single-subject design studies in order to begin building a basis for causal inference. That documentation provides additional information on how to conduct rigorous single-subject research studies for publication, and we recommend close reading of that resource for more information.

Conducting Meta-Analysis With *SSD for R*

SSD for R provides two options for conducting a meta-analysis. The first uses traditional effect size calculations, while the second uses a non-overlap effect size method. The specific method you use is dependent on both the type of data you have and whether you want to include control variables in your analysis. Since non-overlap methods have been traditionally used in single-subject research designs, this method may be preferable if you do not need to include a control variable in your evaluation. Additionally, the method we present here, the mean NAP, can be used in cases where data are more highly autocorrelated and/or significantly trending (Manolov et al., 2011).

Mean Non-overlap of All Pairs

Using the average Non-overlap of All Pairs (NAP) across studies is perhaps the most flexible method to use in the meta-analysis of single-subject research studies. Since trending and serial dependence are not as problematic with the NAP, this would be the preferred method to use if these were issues in any of the studies to be included in your meta-analysis. Therefore, we present this method first.

To begin, you will need to create a file that saves the NAP results for all cases you want included in your analysis. This was described in detail in Chapter 4.

As an example, we have created a file *Student NAP Interventions*. The file contains information about special education students' ability to remain on task prior to and during an intervention, which included reconfiguring classroom space and praising on-task behavior. Data were recorded for seven students, and effect sizes, as noted by NAP values, were calculated and saved in this file. Load these data by entering the following in the Console and navigating to the file location:

```
>Getcsv()
```

Be sure to attach the file before proceeding.

To calculate the average effect size, enter the following in the Console to get the results displayed in Figures 7.1 and 7.2 (see p. 146):

```
> meanNAP(napES, Label, "Effect Size for 7 Special Ed Students")
-----------mean-------------
[1] 0.7891156
------------SD--------------
[1] 0.194144
-------------------------------------------
.93 or above = very effective
.66 to .92 = moderate effectiveness
 below .66 = not effective
-------------------------------------------
```

Figure 7.1 Console output results of **meanNAP()** function

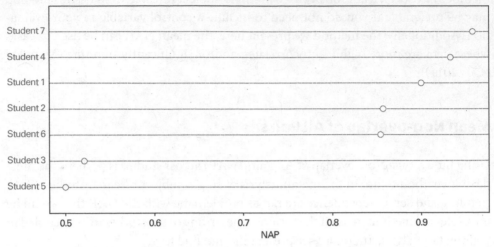

Figure 7.2 Visual output results of **meanNAP()** function

> **meanNAP(napES, Label, "Effect Size for 7 Special Ed Students")**

When entering this command, you will always use **napES** and **Label** as the first two parameters you enter unless you manually change these in your .csv file prior to loading it. Figure 7.1 demonstrates that the mean NAP for all seven students is 0.79 (standard deviation [sd] = 0.19). Looking at the key below the computed output provides information on the interpretation of the calculated values. An average effect of 0.79 is understood to be a moderately effective degree of change.

Figure 7.2 provides additional information. The output in this graph orders studies (in this case students) from lowest effect size at the base of the *y*-axis to highest at the top of *y*-axis. We can see here that Students 3 and 5 did not really improve since their NAP scores fell below the 0.67 mark, which is the threshold for minimal effectiveness. Their NAP scores were much less than the remainder of students included in the study, all of whose effect sizes were greater than 0.8. This is valuable information, and if we were the teachers in this classroom, we might wonder in what ways Students 3 and 5 might differ from those who saw greater effects. This could provide us the beginnings of insights on how we might want to change interventions based on characteristics of more than one study participant.

Using Traditional Effect Size Statistics to Conduct Meta-Analysis

In Chapter 4, we discussed traditional effect sizes such as ES, d-index, and Hedge's *g*. If these effect sizes can be used across studies to be included in a meta-analysis, it may be desirable to do so. Reasons for this, according to Shadish and colleagues (2014), include being able to integrate these studies more easily into comparisons with between-subject designs and making use of well-known meta-analytic tools, such as forest plots and publication bias analyses. Additionally, tools provided in *SSD for R* allow for the introduction of a moderating variable in this analysis.

To illustrate, consider a child welfare agency that provides services to parents whose children are at risk for foster care placement due to allegations of neglect. The agency has recently developed an in-home intervention for which they want to receive a grant from the state. The state, however, will only consider funding the program if it can be shown that participation is effective at keeping families intact. The agency tracked 14 clients who participated in the program using observational measures of parenting before and during the intervention. Effect sizes have been computed for all 14 using the techniques described in Chapter 4, and they have been placed in the file *Parenting intervention meta.csv* that comes with this text.

Use the **Getcsv()** and **attach(ssd)** functions to load these data. If you click on the spreadsheet icon, you will see that there have been two additional columns added to what was developed by saving and appending the effect size file you learned how to create previously. The first is a variable called *Support*, which assess the level of social support available to the parent on a scale ranging from 1 (with no social supports available to the parent) to a high of 10 (the parent has many supports available to them). Additionally, the variable *Single* denotes whether or not the parent has a live-in partner. Zero indicates having a partner, and one indicates being single.

Enter the following in the Console to view the mean effect size across all studies, the results of which are displayed in Figures 7.3 and 7.4 (see p. 148):

>**meanES(ES, Label, "Mean Parenting Skills")**

When entering this command, you will always use **ES** and **Label** as the first two parameters you enter unless you manually change these in your .csv file prior to loading it.

```
> meanES(ES, Label, "Mean Parenting Skills")
-----------mean--------------
[1] 1.326724
-----------SD--------------
[1] 0.9890294
-----------% change--------------
[1] "% change=" "40.77"
************************************************************
small effect size: <.87
medium effect size: .87 to 2.67
large effect size: >2.67
```

Figure 7.3 Console output results of **meanES()** function

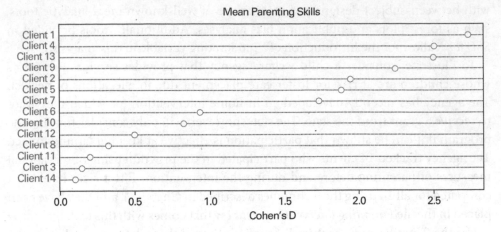

Figure 7.4 Visual output results of **meanES()** function

The output shown in these figures looks similar, although not identical, to what you saw previously. The mean effect calculated is 1.33 (sd = 0.99) and, according to the key provided underneath this output, demonstrates a medium effect since this value is between 0.87 and 2.67. Additionally, we see that, on average, parents saw a 40.77% improvement in parenting skill scores after the introduction of the intervention.

Figure 7.4 shows the effect sizes from smallest to largest. Here, we see that Clients 1, 4, and 13 showed (in decreasing order) the most improvement, while Clients 12, 8, 11, 3, and 14 showed (in decreasing order) the least. Evidence of overall improvement in parenting skills makes a compelling case for potential funders who want some assurances that services they are paying for will actually help clients keep their children at home.

To continue our analysis, you will want to conduct a more robust investigation by using the **metareg()** function as it weights each study based on the variance of the ES within each study. This function assumes that the variance between studies is greater than zero, and that studies included in the meta-analysis are a sample drawn from a larger universe. Therefore, we assume that differences observed between studies is not due to sampling error, but from real differences in the population. The information gleaned from this function will be more substantial and useful. Begin by entering the following in the Console:

> **metareg(ES, V)**

The output for this function is displayed in Figures 7.5 (see p. 149) and 7.6 (see p. 150). Figure 7.5 shows that the summary ES for these 14 studies is 0.94 and is displayed as the *estimate*. The interpretation of this estimate is based on Cohen's guidelines, with 0.2 indicating a small effect, 0.5 indicating a medium effect, and 0.8 indicating a large effect. A summary ES of 0.94 indicates nearly a 1-SD improvement

```
> metareg(ES, V)

  Model Results:

            estimate    se      z  ci.l  ci.u p
    intrcpt    0.940 0.256  3.675 0.439 1.442 0

  Heterogeneity & Fit:

            QE   QE.df    QEp     QM QM.df QMp
    [1,] 23.008 13.000  0.042 13.508 1.000   0

          estimate  ci.lb   ci.ub
    tau^2   0.3698 0.0000  1.8308
    tau     0.6081 0.0000  1.3531
    I^2(%) 48.4951 0.0000 82.3374
    H^2     1.9416 1.0000  5.6617

  Warning message:
  In model.matrix.default(terms, mf, contrasts) :
    non-list contrasts argument ignored
```

Figure 7.5 Console output results of **metareg**() function

from baseline to intervention, which is considered a large effect. The standard error for the ES, the z statistic, as well as the upper and lower bounds of the 95% confidence interval are the additional information provided on this line. The p that is displayed is the p value for the summary statistic. In this case, $p < .05$, so it is statistically significant.

The output from the **metareg**() function also gives us information about the heterogeneity, or the variability, of the studies included in this analysis. The **QEp** value is the Q statistic that tells us whether the heterogeneity between the studies is statistically significant.

The forest plot displayed in Figure 7.6 (see p. 150) illustrates this heterogeneity, or variability, across these studies. Here, we can see that some studies show a very small effect that is not statistically significant (Studies 14 and 3, for instance), while others show much larger effects that are statistically significant but may have wide confidence intervals (Studies 1 and 9, for example). Notice that some studies' ESs (such as Studies 3 and 14) are represented by larger squares, while others are represented by smaller squares. The size of the squares denotes the precision of the ES, with larger samples having more precision. The diamond at the bottom of the forest plot denotes the summary effect (0.940), and the length of it represents the 95% confidence interval, ranging from 0.439 to 1.442.

Since heterogeneity is significant in this case, we should look more closely at the output to determine what might be contributing to the observed variation. Heterogeneity may arise due to sampling error within individual studies. Alternatively, there may be true variation in ESs between studies.

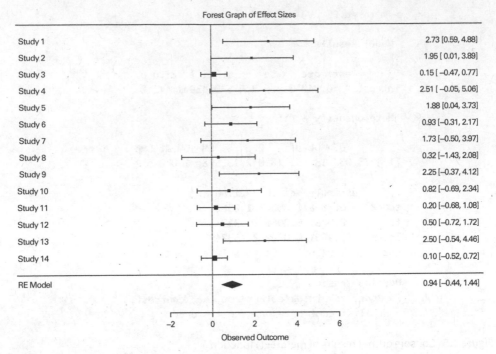

Figure 7.6 Visual output results of **metareg()** function

Previously in our analysis, we used the Q statistic to determine that there was heterogeneity in our meta-analysis, but we can use I^2 to determine both the extent and source of that variation. Values close to 0 indicate that all heterogeneity is associated with within-study sampling error, while values of 100 indicate that all heterogeneity is associated with true between-study differences. The *estimate* for I^2 provides us with information about the magnitude of the observed heterogeneity, with 25% representing low heterogeneity, 50% indicating a moderate amount, and 75% representing a high level of heterogeneity. The I^2 for our example is 48.4951%, indicating a moderate amount of heterogeneity.

The 95% confidence interval provided for I^2 may help us understand the source of the observed heterogeneity. In this example, the 95% confidence interval is wide, ranging from 0 to 82.3374. This illustrates a wide degree of uncertainty in the source of the variation, perhaps due to the small number of studies included in the meta-analysis.

Since there is significant moderate heterogeneity for unknown reasons in this example, it might be helpful to consider moderator variables. We can consider the influence of social support since, theoretically, parents with good social support often have an easier job parenting. Enter the following into the Console to get the output displayed in Figures 7.7 (see p. 151) and 7.8 (see p. 151):

>**metaregi(ES, Support, V)**

```
> metaregi(ES, Support, V)

Model Results:

          estimate     se       z    ci.l    ci.u      p
intrcpt     -0.217  0.234  -0.928  -0.675   0.241  0.353
mods         0.272  0.058   4.679   0.158   0.386  0.000


Heterogeneity & Fit:

          QE   QE.df     QEp      QM  QM.df  QMp
[1,]   1.117  12.000   1.000  21.891  1.000    0


        estimate    ci.lb    ci.ub
tau^2    0.0000  <0.0000  <0.0000
tau      0.0000  <0.0000  <0.0000
I^2(%)   0.0000  <0.0000  <0.0000
H^2      1.0000  <1.0000  <1.0000

The upper and lower CI bounds for tau^2 both fall below 0.
The CIs are therefore equal to the null/empty set.

Warning message:
In model.matrix.default(terms, mf, contrasts) :
  non-list contrasts argument ignored
```

Figure 7.7 Console output results of **metaregi**() function for *support* as a moderator

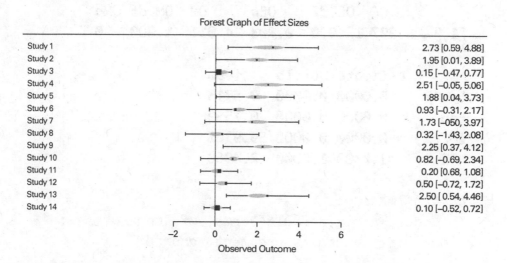

Figure 7.8 Visual output results of **metaregi**() function for *support* as a moderator

In the output shown in Figure 7.7, notice that there is a new line, *mods*, that includes any moderator variables added to the model. The intercept in this case, however, is not significant since $p = .353$ and is greater than the .05 threshold. We therefore ignore this coefficient and discard the model and conclude, in this case, that social support does not improve our understanding of the ESs of the included studies.

Since heterogeneity has not changed or improved, we can think about the variable *Single* since single parents often have a more difficult time parenting. The output displayed in Figures 7.9 and 7.10 (see p. 153) is result of the following being entered in the Console:

>**metaregi(ES, Single, V)**

At first glance, we see that the intercept is significant, with p being displayed as 0, so we can consider this model more closely. Recall that *Single* is a dichotomous

```
> metaregi(ES, Single, V)

  Model Results:

            estimate      se      z   ci.l   ci.u p
  intrcpt      2.213   0.464  4.773  1.304  3.121 0
  mods        -1.848   0.494 -3.743 -2.815 -0.880 0


  Heterogeneity & Fit:

            QE   QE.df     QEp      QM  QM.df QMp
  [1,]   8.997  12.000   0.703  14.011  1.000   0


          estimate  ci.lb   ci.ub
  tau^2     0.0000 0.0000  0.5715
  tau       0.0000 0.0000  0.7559
  I^2(%)    0.0000 0.0000 59.9365
  H^2       1.0000 1.0000  2.4960

  Warning message:
  In model.matrix.default(terms, mf, contrasts) :
    non-list contrasts argument ignored
```

Figure 7.9 Console output results of **metaregi()** function for *single* as a moderator

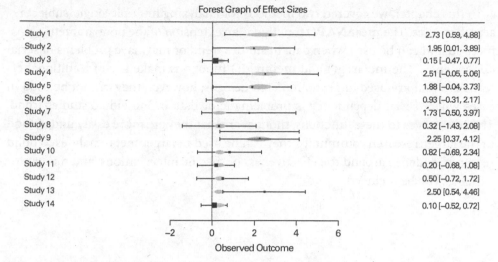

Figure 7.10 Visual output results of **metaregi**() function for *single* as a moderator

variable, so the interpretation of the output with this moderator is slightly different from that of continuous variables. The value of the intercept is 2.213, which is the ES when *Single* = 0 (i.e., being partnered). The *p* value for the moderator is also significant, so we can understand that as the slope. That is, when going from 0 to 1 (i.e., partnered to single), there is a 1.848 reduction in parenting skills. Therefore, the effect size for single parents is 2.213 −1.848 = 0.365. It appears as if improvement in parenting skills is moderated by parents being single or partnered.

Also note that heterogeneity is no longer significant since **QEp** is no longer significant, and I^2 is estimated at 0. Therefore, the inclusion of *Single* in the meta-analysis improves our understanding of the ES of the parenting intervention.

Conclusion

Being able to combine single-subject research designs is particularly useful in both practice and research settings. From a practice perspective, it provides more information about what practitioners might be able to expect from clients as a result of implementing specific interventions. In the examples provided, we were able to identify clients who responded to interventions better than others. We also found that being single negatively impacted parenting skills across studies. In this way, we might be able to begin identifying the types of clients who might respond more successfully to specific ways of working than others. From a research perspective, being able to conduct meta-analyses of single-subject research designs adds to the knowledge base of what is known about particular interventions by providing rigor that is not attainable by analyzing one study at a time.

In this chapter, we covered two methods for analyzing multiple single-subject research studies. The **meanNAP()** function is an extension of the nonparametric NAP method and can be used when data in studies trend or may have problems of serial dependency. The **metareg()** and **metaregi()** functions make use of traditional effect sizes that are used in group research designs; however, they cannot be used in cases where serial dependency is problematic or data in individual studies trend. The advantages to these functions, though, are that they are more easily understood in the larger research community, they can be used in larger meta-analyses to build research evidence around the effectiveness of specific interventions, and moderator variables can be included.

8
Using *RMarkdown* to Present Your Findings

Introduction

In this chapter you will learn how to utilize *RMarkdown* to present *SSD for R* findings in a well-ordered and reproducible manner. *RMarkdown* is a plain text formatting syntax that makes writing research reports simple. The language provides a simple syntax that formats text such as headers, lists, boldface, and so on. This language is popular, and you will find many apps that are compatible with it. For example, combined with other packages, like *SSD for R*, users can easily create tables and graphics to present their research findings. Another important feature of this markdown language is that it will make your findings reproducible in that all of your files are connected. Thus, if there are changes to your data, rerunning the analysis is simple. As Baumer and Udwin (2015) suggested, an *RMarkdown* document links computation, output, and written analysis to enhance transparency, clarity, and ease to reproduce the research. Furthermore, sharing data is only a click away (Baumer & Udwin, 2015).

As mentioned, *RMarkdown* can save and execute *SSD for R* code and create high-quality reports from a single file. Doing so provides greater insight into your results and clearly displays what you did to analyze your data. Furthermore, *RMarkdown* makes it possible for other clinicians/researchers to replicate findings. *RMarkdown* is a straightforward language that allows you to create documents with headings, text, images, margins, and more to enhance the communication of the findings produced by *SSD for R*. Documents created in *RMarkdown* can be saved in different file formats, such as doc, PDF, or HTML.

Getting Started

If you are using a stand-alone version of *RStudio*, the installation of a LaTex package is required to create professional-looking files in PDF format. LaTeX is already installed on *RStudio* Cloud.

To create a Word file, the installation of MS Word is required. For a Windows computer, MiKTeX is recommended and can be installed from the following URL: http://miktex.org/download. For a Mac, MacTex 2013 + is recommended and can be installed from the following URL: https://tug.org/mactex/mactex-download.html. When you create a report utilizing *RMarkdown*, it will automatically convert the report to a PDF or Word file without your direct interaction with the LaTeX software you installed. Once again, LaTeX is already installed on *RStudio* Cloud.

SSD for R. Charles Auerbach and Wendy Zeitlin, Oxford University Press. © Oxford University Press 2022.
DOI: 10.1093/oso/9780197582756.003.0009

The next step is to open *RStudio* or *RStudio* Cloud and type the following commands in the Console one at a time to install the necessary packages to create a report in PDF or Word format:

```
>install.packages("rmarkdown")
>install.packages("knitr")
>install.packages("kableExtra")
```

Creating an *RMarkdown* Document

One of the primary reasons for developing an *RMarkdown* file is its ability to be reproduced by other clinicians and researchers. As a result, interactive commands such as **ABlines()** in *SSD for R* will not run. These commands have been replaced and begin with the prefix RM; for example, **RMlines()** is substituted for **ABlines()** to draw a vertical line between phases. The table that follows displays a list of all interactive commands that have substitutes for writing *RMarkdown* scripts. The best way to obtain the coordinates for the commands that follow is to create the graph in *RStudio* and then apply them to the commands in the chunk.

Interactive Command	*RMarkdown* Substitute	Syntax	Example
ABarrow()	**RMarrow()**	RMarrow(X1,Y1,X2,Y2) X1 y1 **From** coordinates of arrow. X2 y2 **To** coordinates of arrow.	>RMarrow (5.9,5,9.5,5.1)
Gline()	**RMGline()**	RMGline(Y) **From** Y coordinate to draw horizontal line	>RMGline(5)
ABlines()	**RMlines()**	RMlines(behavior, X) Behavior variable to draw vertical line **From** X coordinate to draw vertical line	>RMlines (yell,15.5)
ABstat	**RMstat()**	RMstat(behavior, phase,"statistic",X) X start of horizontal line	>RMstat (yell,pyell,"A", "mean",.1)
ABtext()	**RMtext()**	RMtext("text",X,Y) **From** X and Y coordinates to draw text	>RMtext ("argument",2.5,5)
Getcsv()	**ssd<-read()**	ssd <- read.csv("file") file data file name	>ssd <- read. csv("Jennyab.csv")

Once this is done, you are ready to work through an example. The example will focus mostly on creating an MS Word document. The first step is to create an *RMarkdown* file (.Rmd). In *RStudio*, select File/New File/R Markdown, and the menu in Figure 8.1 will be displayed. Select Create Empty Document and a blank untitled *RMarkdown* document will appear in the top left pane.

The first step is to create a YAML where you can configure your *RMarkdown* file. This portion contains the header, author, title, and type of format you desire. This portion of the file starts and ends with ---. Here is an example:

```
---
title: "SSD for R"
author:
- Charles Auerbach
- Wendy Zeitlin
subtitle: An R Package for Analyzing Single-Subject Data
output:
word_document: default
pdf_document: default
---
```

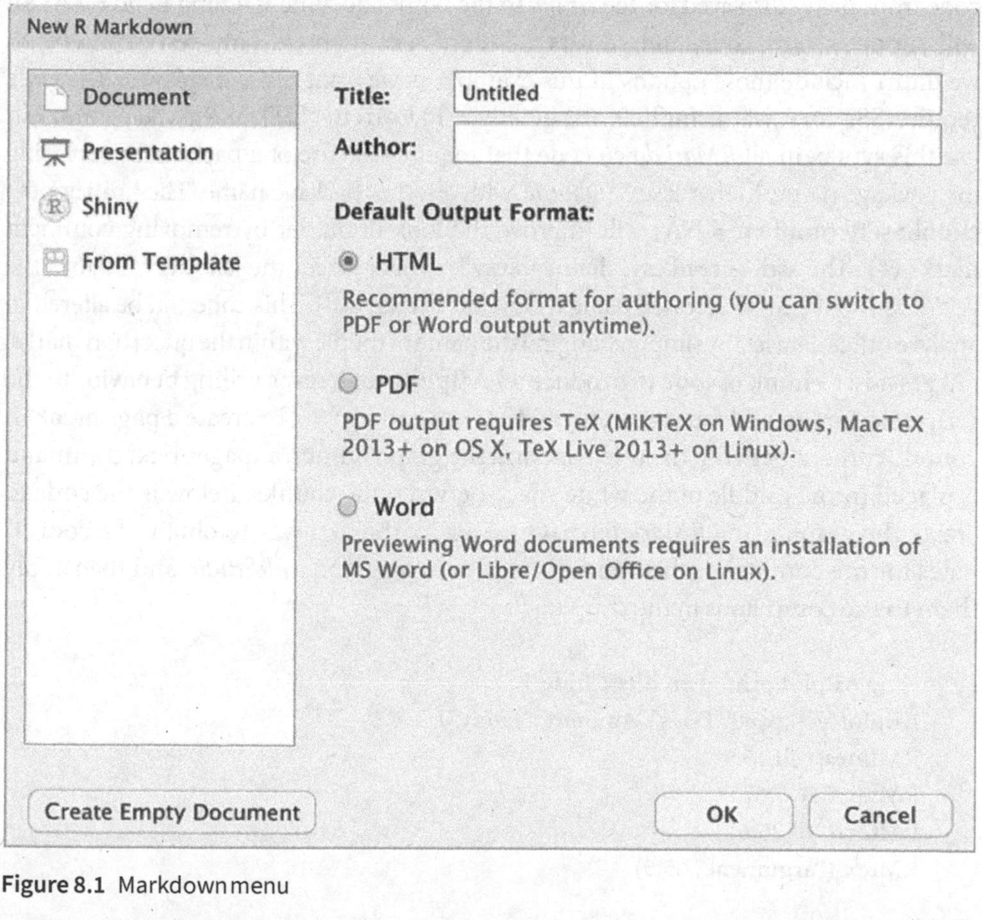

New R Markdown

Document	**Title:** Untitled
Presentation	**Author:**
Shiny	**Default Output Format:**
From Template	

HTML
Recommended format for authoring (you can switch to PDF or Word output anytime).

PDF
PDF output requires TeX (MiKTeX on Windows, MacTeX 2013+ on OS X, TeX Live 2013+ on Linux).

Word
Previewing Word documents requires an installation of MS Word (or Libre/Open Office on Linux).

Create Empty Document   OK   Cancel

Figure 8.1 Markdown menu

In this case, there are two authors, and each author's name begins with a hyphen (-) separated by a space. Also, note that under the output section, both Word and PDF documents are entered, allowing for substituting between document types.

Below the YAML is where different *code chunks* begin, which contain various *R* functions from *SSD for R*. For example, the *code chunk* below will open the *jennyab.csv* file, which can be downloaded from the authors' website (https://www.ssdanalysis.com) or the publisher's website. The data file must be placed in the same directory as the *.rmd* file.

```
```{r openfile,include=FALSE}
knitr::opts_chunk$set(comment = NA)
require(SSDforR,warn.conflicts=F, quietly=T)
ssd <- read.csv("Jennyab.csv")
attach(ssd)
```
```

Each *code chunk* begins and ends with three ``` (grave accents). Inside the {} curly brackets, the *r* indicates the chunk will include *R* code. Each chunk requires a unique name, and in this example, the chunk was given the name *openfile*. After the comma (,) options can be added for the chunk. Using the option **include = FALSE** will suppress code from being displayed for the *chunk* in the output document. Using **Echo = FALSE** will suppress only code, and **results = "hide"** will suppress only results; however, we didn't include those options in this example as we want those displayed. The code **require(SSDforR,warn.conflicts = F, quietly = T)** loads the *SSD for R* package. You can use this syntax in all *RMarkdown* code that requires the use of a package by replacing the package name, in this case *SSD for R*, with another package name. The **knitr::opts_chunk$set(comment = NA)** will improve the look of output by removing comment marks (#). The **ssd <- read.csv("Jennyab.csv")** syntax places the *jenny.csv* file into the vector *ssd* so it can be accessed using the syntax **attach(ssd)**. This code can be altered to analyze other datasets by simply changing the name of the file within the quotation marks.

Let's add a chunk of code to produce an **ABplot()** of Jenny's yelling behavior to the script. Be sure to add four line returns after the ending ```. To create a page break in your document between the title page and the graph output, a **pagebreak** command is placed in the middle of the white space between the chunks. Below is the code to create the graph using *RMarkdown*. Once again, the best way to obtain the coordinates for the commands that follow is to create the graph in *RStudio* and then apply them to the commands in the chunk.

```
```{r ABplot,echo=F,results="hide"}
ABplot(yell,pyell,"Days","Amount","Jenny")
RMlines(yell,15.5)
RMtext("A",5,6)
RMtext("B",25,6)
RMtext("argument",2.5,5)
```
```

```
RMarrow(5.9,5,9.5,5.1)
RMstat(yell,pyell,"A","mean",.1)
```
```

Once again, the chunk begins with a title and the following options: **echo = F** and **results="hide"**. The **echo** is set to "F", and the **result** options are set to "**hide**", resulting in the prevention of messages and results from being printed in the document. The *SSD for R* **ABplot()** produced will appear in the output document. The function that follows will annotate the graph. The best way to obtain the coordinates for the commands that follow is to create the graph in *RStudio* and then apply them to the commands in the chunk. The **RMlines()** function will place a vertical line on the *x* ordinate 15.5; the **RMtext()** will place an "A" at the *x* ordinate 5 and the *y* ordinate 6; the second **RMtext()** function will place a "B" at the *x* ordinate 25 and the *y* ordinate 6. The final **RMtext()** function will place the text "argument" at the *x* ordinate 2.5 and *y* ordinate 5; the **RMarrow()** will draw an arrow from the text "argument" to the 10th observation. Finally, the **RMstat()** function places a mean line in the baseline.

To create the document, we can *knit* the .rmd file to create the MS Word document to view the graph. When you knit a file, your *RMarkdown* document is automatically saved. In order to do so, Word must be installed on your computer. As displayed in Figure 8.2, in the top left pane click on the down arrow next to knit and select MS Word.

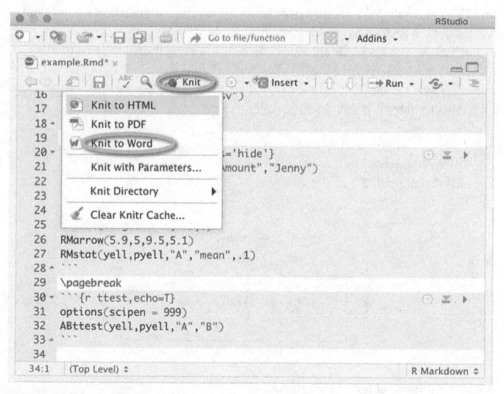

**Figure 8.2** Knit menu for RMarkdown
```

The output will be displayed in a Word file. If the file is read only, use File/Save As to save the file under a new name. If you are running *RStudio* Cloud, when the output is created, a message will appear to downloaded the file to your local machine. You will then be able to modify the file. Figure 8.3 displays the result that will be produced in a Word document.

You can also add text to annotate your findings. For example, the following can be included by pressing return four times after the ending ``` .

Introduction

The School Base Support team recommended that Jenny receive Cognitive Behavioral Therapy (CBT) at the Child Help Center to reduce or eliminate maladaptive and inappropriate behaviors. The primary goals were the development of self-control and problem-solving strategies. The acquisition and internalization of these skills provide the means for the child to regulate her behavior.

Visual Analysis

For data analysis, visual analysis was used. Results indicated that the trend of changes in the behavior scores for the client was descending in the interventions phase (B) and indicated improvement. Figure 1 illustrates the results from the AB design.

\pagebreak

Note the double asterisks (*) around the subtitle to bold the subtitles "Introduction" and "Visual Analysis." You add a page break by placing the **\pagebreak** command as a line after the final sentence of the section's explanation.

Each phase was tested for autocorrelation by inserting the *chunks* that follow:

```
```{r autoA,echo=T}
ABrf2(yell,pyell,"A")
```
```

```
```{r autoB,echo=T}
ABrf2(yell,pyell,"B")
```
```

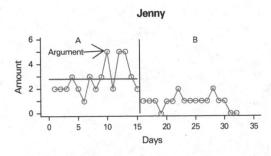

Figure 8.3 ABplot() of Jenny's crying behavior

```
ABrf2(yell,pyell,"A")

[1] "tf2="  "0.801"
[1] "rf2="  "0.267"
[1] "sig of rf2=" "0.432"
----------regression-----------

Call:
lm(formula = A ~ x1)

Coefficients:
(Intercept)          x1
    1.7429       0.1321
```

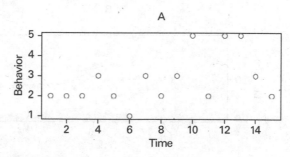

Figure 8.4 Baseline **ABrf2**() results

Notice that once again, the option **echo** was set to T to display the *SSD for R* code in the document. To view the results of the *autoA* and *autoB* chunks, you will need to knit the *.rmd* file. The results are shown in Figures 8.4 and 8.5 in the resulting output from the autocorrelation *chunk* for the baseline autocorrelation. Figures 8.6 (see p. 162) and 8.7 (see p. 162) display the results for the intervention autocorrelation.

The next chunk will produce the results of a *t* test comparing the A and B phases. Note the title of the chunk is **ttest** and the option **echo = T** is added to include the syntax in the output. The syntax **options(scipen = 999)** is added to remove scientific notation from the output. The *SSD for R* syntax **ABttest(yell,pyell,"A","B")** performs the *t* test.

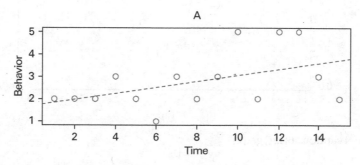

Figure 8.5 Baseline **ABrf2**() Graph

```
ABrf2(yell,pyell,"B")

[1] "tf2="    "1.294"
[1] "rf2="    "0.383"
[1] "sig of rf2=" "0.209"
-----------regression-----------

Call:
lm(formula = A ~ x1)

Coefficients:
(Intercept)          x1
    1.11765     -0.01961
```

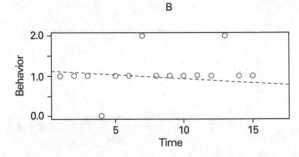

Figure 8.6 Intervention **ABrf2()** results

```
```{r ttest,echo=T}
options(scipen = 999)
ABttest(yell,pyell,"A","B")
title(main= "Figure 2: Mean Differences")
```
```

Once again, *knit* the *.rmd* file as displayed in Figure 8.2, and changes made to the file will be automatically saved. Figures 8.8 (see p. 163) and 8.9 (see p. 164) display the result of the *t* test. Note because **echo** = T was entered as an option, the syntax is also

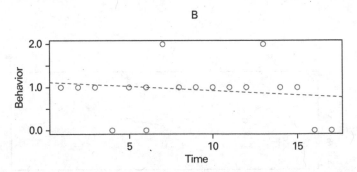

Figure 8.7 Intervention **ABrf2()** graph

```
options(scipen = 999)
ABttest(yell,pyell,"A","B")
```

```
	Two Sample t-test

data:  A and B
t = 5.4965, df = 30, p-value = 0.000005735
alternative hypothesis: true difference in means is not equal to 0
95 percent confidence interval:
 1.168160 2.549487
sample estimates:
mean of x mean of y
2.8000000 0.9411765


	F test to compare two variances

data:  A and B
F = 5.181, num df = 14, denom df = 16, p-value = 0.002407
alternative hypothesis: true ratio of variances is not equal to 1
95 percent confidence interval:
   1.839162 15.145963
sample estimates:
ratio of variances
         5.180952


	Welch Two Sample t-test

data:  A and B
t = 5.2611, df = 18.7, p-value = 0.00004685
alternative hypothesis: true difference in means is not equal to 0
95 percent confidence interval:
 1.118519 2.599128
sample estimates:
mean of x mean of y
2.8000000 0.9411765
```

Figure 8.8 Results from **ABttest()** function

included in the document. The **title()** function is included to provide a title for the mean bar plot produced by the **ABttest()** function. The **title()** function adds the title inside the quotations to the bar graph produced with the **ABttest()** function. Notice that "t" in t-test is between asterisks, which will italicize the characters between them.

The final *chunk* will add the results of an effect size. Once again *knit* the *.rmd* file as displayed in Figure 8.2, and changes made to the file will be automatically saved. Figure 8.7 displays the result of the effect size.

```
```{r ES,echo=T}
Effectsize(yell,pyell,"A","B")
```
```

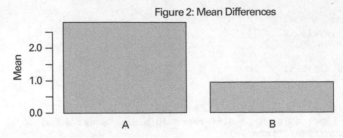

Figure 8.9 Graph from **ABttest()** function

Again, you can also add text to annotate your findings and conclusions. For example, the following can be included by adding four white spaces after the ending ```.

> **Statistical Analysis**
> Both phases were tested for autocorrelation. The findings indicate that the degree of autocorrelation was small enough in each phase (rf2 = 0.267 and rf2 = 0.383, respectively) to conduct a *t*-test for statistically significant differences between phases. A Welch correction was used to compensate for unequal variances. The mean number of yelling episodes per day decreased from 2.8 to .94 from baseline to intervention. Figure 2 shows these differences, which are statistically significant (*t* = 5.26, *p* = <001). The Cohen's *d* of 1.95 indicates a moderate degree of change between phases.
>
> **Conclusion**
> The School Based support team is proposing removing the intervention to form an A-B-A design, which is sometimes referred to as an experimental removal of intervention design. The design is experimental because it consists of testing whether the introduction of an intervention was likely to have caused changes in the target behavior as a result of the child's integration of problem-solving skills to regulate her behavior.

To view the results of *ES chunk* along with the additional narrative, *knit* the .rmd file once again. The results are shown in Figure 8.10 (see p. 165). If the MS Word document is read only, to edit the document, it will need to be saved under a new file name.

Conclusion

In conclusion, this chapter provided an introduction to developing *RMarkdown* documents to effectively present your *SSD for R* findings. There are a number of helpful texts and websites on using *RMarkdown* in Appendix D. These examples provide a more in-depth explanation on formatting text, tables, and figures.

```
Effectsize(yell,pyell,"A","B")

small effect size: <.87
medium effect size: .87 to 2.67
large effect size: >2.67
*************************************************************
***********************ES**********************************
                       B                              B
"ES=         "       "1.46953"   "% change="       "-42.92"

        lower  effect    upper
B 0.5742522 1.46953 2.332266
*****************d-index***********************************
                       A                              A
"d-index=    "       "1.94711"   "% change="       "47.42"

        lower  effect    upper
A 1.086059 1.947109 2.786267
*****************Hedges's g********************************
                       A                              A
"Hedges's g="        "1.89795"   "% change="       "47.11"

        lower  effect    upper
A 1.044087 1.89795 2.73013
*****************Pearson's r******************************
[1] 0.708
*****************R-squared********************************
[1] 0.502
(s)ave, (a)ppend, or (n)either results? (s/a or n)
```

Figure 8.10 Reults from **Effetsize()** function

Chapter Exercise

Assignment 8.1—Create an *RMarkdown* Report on Evaluating Your Practice With Brenda

For this assignment, you will use Brenda's baseline and intervention data to create a final report describing your practice evaluation. You will utilize *RMarkdown* to present *SSD for R* findings of the Brenda case example in a well-ordered and reproducible manner. You can modify the *RMarkdown* example included with the text to create your document.

1. Create a YAML header.
2. Create a chunk *load SSDforR*, open the Brenda baseline/intervention file and attach it.

3. Create a chunk to recreate an **ABplot()** you created in Assignment 4.1.
4. Add an Introduction
5. Create a chunk to display your findings from Assignment 4.3.
6. Create a chunk to display the results on the effect size you utilized in Assignment 4.3.
7. Create a chunk to display the appropriate hypothesis test used in Assignment 5.3. Also add a description of your statistical analysis.
8. Finally add a conclusion.

9
Building Support for Practice Research

Introduction

We recognize that practice research does not occur in a vacuum, and those desiring to do this type of evaluation work in many different types of settings. Some practitioners work in solo practices, while others may be employed in large multisite, multiprogram organizations with thousands of employees and clients at any given time. Therefore, the notion of engaging in practice research and the ability to do so should be considered within the context in which one works, as flexibility, resources, and work demands can vary from organization to organization.

In this chapter, we discuss common issues that arise when organizations, whether small or large, wish to engage in practice research, and we provide some suggestions for addressing these.

It has been widely recognized that engaging in practice research is an essential component of good and responsible practice in many disciplines (Shaw, 2011). While we discussed the benefits of this type of research previously, practice research findings can also be used at an administrative level to improve existing programs and plan for future services. Engaging in practice-based research can also be used to develop organizational policies and procedures and aid in decision-making about the allocation of resources (Preskill & Boyle, 2008).

We believe, then, that the importance of practice research cannot be overstated. Despite this, many wishing to engage in practice research often face challenges, including issues related to the practical ability to conduct research, often referred to as research capacity.

Research Capacity and Research Culture

Building research capacity "attempts to enhance individual and collective research expertise and hence capacities to contribute to improving the quality and quantity of . . . research" (Christie & Menter, 2009, p. 5). Another description of research capacity includes the ability and motivation to routinely conduct evaluations of practice (Preskill & Boyle, 2008). Factors that contribute to engagement in practice research include abilities and desire. Therefore, our discussion on building support for practice research in this chapter focuses on both capacity and organizational culture.

SSD for R. Charles Auerbach and Wendy Zeitlin, Oxford University Press. © Oxford University Press 2022.
DOI: 10.1093/oso/9780197582756.003.0010

Barriers to Conducting Research

It is necessary to consider the barriers facing those wishing to engage in practice research, as identifying and addressing these will help organizations build research capacity. Barriers to conducting research within organizations generally fall into two broad categories: organizational challenges and individual factors. Organizational challenges include administrative issues and the role of practitioners within the organization. Individual factors contributing to challenges with conducting practice research include skills and the nature of the profession to which the practitioner belongs.

Organizational Challenges to Conducting Research

Frequently, there may be an overall desire to engage in practice research, but there are challenges that must first be overcome. For example, most human service organizations wishing to evaluate their practices are not primarily research institutions; they are service oriented. In these organizations, conducting research may not be a priority, there may be negative attitudes about research in general, there may not be administrative support for quality improvement, or there may be an overall lack of support for research activities (Beddoe, 2011; Epstein & Blumenfield, 2012; Fouché & Lunt, 2010; McCrystal & Wilson, 2009). One reason given for this lack of focus on research activities is management being unsure of what to do with research findings and the value these activities could bring to its core mission (Beddoe, 2011).

In many cases, there may be desire among individual practitioners or administrators to evaluate practice, but an absence of management support to engage in these activities can lead to substantial roadblocks. For instance, organizations focused on service delivery may not have the resources readily available to conduct research. This could include difficulty accessing professional development that supports practice research, computers, technology support, and/or scholarly literature (Beddoe, 2011; Carman & Fredericks, 2010; Lunt et al., 2008).

In addition to administrative issues, another type of organizational barrier to conducting practice research revolves around the role of practitioners within organizations. A primary concern for practitioners wishing to engage in practice research is time. Many practitioners working in human services have heavy caseloads with barely enough time to complete the requirements of their daily work. Insufficient time to conduct practice research is the reason most often cited for not doing so (Beddoe, 2011; Carman & Fredericks, 2010; Edwards & Cable, 2009; Epstein & Blumenfield, 2012; Fouché & Lunt, 2010; Lunt et al., 2008; McCrystal & Wilson, 2009).

Individual Barriers to Conducting Research

The other most commonly cited reasons for not engaging in practice research are related to ability. In many cases, practitioners lack the confidence, knowledge, or skills to engage in practice research (Beddoe, 2011; Carman & Fredericks, 2010; Edwards & Cable, 2009; Epstein & Blumenfield, 2012; Fouché & Lunt, 2010; Lunt et al., 2008; McCrystal & Wilson, 2009). Specifically, practitioners are often unsure of what to measure or how to measure client problems or their progress (Carman & Fredericks, 2010). They may have difficulty managing data and using statistical software (Carman & Fredericks, 2010). Finally, many practitioners are not sure how to utilize or effectively disseminate research findings in general (Edwards & Cable, 2009).

A related systemic problem surrounding practitioner research skills has to do with the nature of the profession to which practitioners belong. For example, while social workers have deemed the development of research skills important, there has traditionally been little focus on this within the discipline itself (Edwards & Cable, 2009; Lunt et al., 2008; McCrystal & Wilson, 2009; J. Orme & Powell, 2008). This has been exhibited by insufficient attention paid to developing research skills in professional programs (Epstein & Blumenfield, 2012; Lunt et al., 2008; McCrystal & Wilson, 2009). This is mirrored in other human services professions as well.

Additionally, there are few postgraduate educational opportunities across professions to teach practitioners about research methods and building research skills (J. Orme & Powell, 2008).

Despite these challenges, and as we referenced in our introductory chapter, we believe that there will be an increasing demand for practice research in the future. The trends toward utilizing evidence-based practices and increasing accountability in healthcare and behavioral healthcare are indicative of the need to build organizational environments conducive to engaging in practice research within all human services organizations.

Developing Research Capacity and a Culture of Research Within Organizations

Action items have been developed and implemented in order to make research activities accessible to practitioners and others wishing to conduct research in human services organizations. These have been aimed at creating environments that encourage research among practitioners while building the infrastructure necessary to actually engage in these activities. Action items fall into two categories—those related to infrastructure and demand and those related to skill building. While we can categorize these, it should be noted that to implement all of these activities requires concerted effort and administrative support.

Developing the Infrastructure and Demand for Research

Building the infrastructure and demand for practice research includes activities related to developing overall infrastructure, supporting research activities, valuing research as an important activity at the organizational level, and producing useful research outputs. While this may initially seem burdensome, it lays the foundation for longer term success; when administration values and supports research, evidences suggests more effective research capacity building within organizations (Arnold, 2006; Beddoe, 2011; McCrystal & Wilson, 2009; J. Orme & Powell, 2008).

A number of suggestions have been made to assist with building research infrastructure specifically. All of these require thoughtful planning, the consideration of integrating practice research into the core activities of the organization, and intentionally building research capacity (Christie & Menter, 2009). One recommendation is to conduct an evaluation of the organization's existing research capacity. Baizerman, Compton, and Hueftle (2002) have developed a checklist to assist in this activity by examining existing processes, practices, and professional capabilities.

Based on this, organizations should consider developing or identifying practice research models that are congruent with the aims of the organization and resources available to them (Carman & Fredericks, 2010). Arnold (2006) further supported this by suggesting that evaluation activities should be considered in human services organizations when programs are initially designed. To assist with this, organizations should think about developing a holistic approach to building research capacity.

Volkov and King (2007) have developed a checklist to help organizations develop research capacity. This checklist, which is referenced in Appendix D, looks at the context, including organizational culture in which research activities occur, mechanisms within the organization that are conducive to building research capacity, and identification of resources necessary to conduct practice research.

Most obviously, building infrastructure should also include providing access to computers, databases, and software to analyze data (Edwards & Cable, 2009; J. Orme & Powell, 2008; Rock et al., 1993).

Another factor to consider in developing organizational research infrastructure involves human resources. For instance, some suggestions have included identifying a champion within the organization who understands both practice and research in order to help integrate practice research into the core activities of the organization. These activities require incremental steps that include being responsive to practitioner needs and work demands (Edwards & Cable, 2009; J. Orme & Powell, 2008; Rock et al., 1993). In fact, practitioners should be included in regular meetings and decision-making at all stages of capacity building in order to develop an integration plan that can pragmatically be absorbed into the culture of the organization (Rock et al., 1993).

Additionally, in order to build infrastructure, organizations should consider supporting doctoral studies and including research and development requirements in job descriptions (J. Orme & Powell, 2008). Not only do these human resource considerations help build staff capabilities, but also they include the expectation that employees will be involved in practice research (J. Orme & Powell, 2008). To build more support for practice research, evaluation findings should be shared with administrators and practitioners in a manner that is both meaningful and useful (Arnold, 2006). Finally, conducting practice research regularly helps incorporate these activities into the awareness and day-to-day work of practitioners (Arnold, 2006).

Building a culture of research occurs within organizations, but it is also influenced by the community stakeholder groups. For example, the notion of valuing the process of research and its outputs in any one organization is supported by others within the same sector, board members, funders, and policymakers (Carman & Fredericks, 2010; Lunt et al., 2008; Shaw, 2011). Therefore, stakeholder groups can work together to build demand for practice research activities at the administrative level. This can result in administrators encouraging practice research and working to increase the visibility of research activities within organizations (McCrystal & Wilson, 2009).

Building Research Skills Within Organizations

In general, building research capacity can be enhanced by the development of pragmatic research skills. This includes developing individual skills and building collaborative relationships that are supportive of research activities.

Practice research skills can be developed among workers. As previously stated, two ways of doing this are by developing formal policy decisions that include supporting doctoral studies and including research skills in job descriptions (J. Orme & Powell, 2008; Volkov & King, 2007). The goal of these should include developing research knowledge and skills in the majority of practitioners (Arnold, 2006).

Organizations should also create and support learning opportunities for employees. This could include providing research training for service users and front-line practitioners (J. Orme & Powell, 2008). Other ways of supporting employees' learning about research methods and skills involves encouraging employees to participate in research internships and to attend online or in-person trainings, workshops, and seminars; and providing access to reading material (Christie & Menter, 2009; Preskill & Boyle, 2008). Finally, organizations can encourage employees who have the capacity to become practice research leaders to participate in train-the-trainer learning opportunities in which participants are not only taught relevant research knowledge, but also are provided the resources and skills to pass these along to others (Edwards & Cable, 2009).

Another way to provide organizations with usable research skills is through collaborative activities. This can involve something as simple as contracting with experienced practice research evaluators to obtain technical assistance and to regularly work with practitioners (Epstein & Blumenfield, 2012; Preskill & Boyle, 2008). One possible venue for accessing these skills is by actively working to create university-organization partnerships not to only conduct evaluations, but also to help develop practice environments that are conducive to ongoing practice research within service organizations (McCrystal & Wilson, 2009; J. Orme & Powell, 2008). A suggestion provided by Shaw (2005) goes one step further and directs universities to consider establishing centers of practice research. In this way, universities can become known and easily identified resources that allow organizations to build research capacity without the burden of having to build their own research collaboratives.

However collaboration is included in practice research, work should begin with teamwork between researchers and consumers of research to determine what outputs are needed (Berger, 2010).

Regardless of how research knowledge is brought into organizations, numerous recommendations have been developed to encourage continuing skill development and ongoing research activities within the organization itself. Rock and colleagues (1993) suggest that including practitioners in research activities and decision-making not only improves attitudes about the research process but also breaks down resistance to this across the organization.

Research collaboration within organizations can take many forms. One of the most commonly written about is research mentorships. In these types of collaborations, more experienced researchers teach practitioners about research methodology and skills by including them in meaningful projects and helping them take responsibility for portions of the research (Christie & Menter, 2009; Edwards & Cable, 2009; Preskill & Boyle, 2008). In this way, one-on-one support can be provided that allows practitioners to gain research skills incrementally (Arnold, 2006). Some variations of research mentorship involve the creation of evaluation teams where more experienced teams provide assistance to less experienced teams through various projects (Carman & Fredericks, 2010). Another way to engage in mentorship is to pair academic researchers with practitioners for co-learning projects. In these, practitioners gain research skills while academics learn about the pragmatics of organizational life (Fouché & Lunt, 2010). Finally, Carman and Fredericks (2010) described research mentoring to involve the creation of learning networks that integrate peer learning and the sharing of best practices. In this way, mentorship is a reciprocal relationship where practitioners and researchers participate in collaborative, mutually beneficial activities over time.

Conclusion

The focus of this book has been on single-subject research design, methodology, and analysis; however, in order to make this, or any type of research, accessible to organizations and practitioners, a clear plan for conducting practice research should be developed.

This chapter was written to give you the opportunity to consider the context of your organization along with what may be required to include practice research into your organization's core activities. This chapter was not meant, however, to be an exhaustive discussion of building research capacity; rather, it was designed to provide readers a broad understanding of factors that should be considered when attempting to implement research in a practice environment.

Common obstacles to conducting research in practice settings discussed in this chapter include administrative factors, work demands placed on practitioners, the availability of research knowledge and skills, and the research tradition of some professions. To address these, recommendations have been developed to remediate these barriers. These involve building support and demand for practice research by increasing its value to stakeholders, the development and/or accessibility to research skills, and providing the infrastructure necessary to conduct practice research.

In this chapter we discussed the importance of including practitioners in the process of building research capacity. It should be emphasized that, in order to increase the chances of success, capacity building must be collaborative. All activities should include representation from all employee groups that will either participate in the research process or will be consumers of research. Organizational change of any type needs to consider the needs of all involved (Choi & Ruona, 2011).

With the publication of this book, we hope that we have begun to address some of the barriers identified in this.

ENTERING AND EDITING DATA DIRECTLY IN *R*

ENTERING DATA DIRECTLY INTO *SSD FOR R*

Throughout the text, we are assuming that you are using Microsoft Excel or a similar program to enter your data; however, there may be situations in which these programs may not be available to you. Fortunately, there is way to enter data directly into *R* for use with the *SSD for R* package. Before beginning it is necessary to open *RStudio*. In the Console, type the following command:

>**require(SSDforR)**.

As an example, let's track two client behaviors for a hypothetical client: amount of crying and level of self-esteem. An A-B-B_1 design is being utilized. The A and B phases each have six observations, while the B_1 phase has eight observations.

Entering data in *R* has to be done in a specific format as displayed below.

>cry<-c(3, 4, 2, 5, 3, 4, NA, 2, 2, 3, 2, 1, 2, NA, 2, 2, 1, 2, 1, 0, 0, 0)
pcry<-c("A", "A", "A", "A", "A", "A", NA, "B", "B", "B", "B", "B", "B", NA, "B1", "B1", "B1", "B1", "B1", "B1", "B1", "B1")
esteem<-c(3, 4, 2, 5, 3, 4, NA, 2, 2, 3, 2, 1, 2, NA, 2, 2, 1, 2, 1, 0, 0, 0)
pesteem<-c("A", "A", "A", "A", "A", "A", NA, "B", "B", "B", "B", "B", "B", NA, "B1", "B1", "B1", "B1", "B1", "B1", "B1", "B1")

Note that the behavior variable *cry* is followed by a "<" (less than sign) and a "–" (dash) which is used synonymously in *R* to an equal sign (=). Also note the lower case "c" before the open parenthesis and that a comma follows each observation entry. Note that for the phase variables, *pcry* and *pesteem* the letters *A*, *B*, and *B1* are enclosed in quotes. Finally, notice that the same number of elements are entered for the behavior variables and their corresponding phase variables. Additionally, the **NA**, inserted as described in Chapter 1 to denote a change of phases, is located in the same position in the behavior variable and its corresponding phase variable. For example, *cry* is made up of a total of 22 elements, as is the corresponding phase variable, *pcry*. We also note that phases change for both the behavior and phase variables after six occurrences of the baseline, and six occurrences of the first intervention phase. After you enter the close parenthesis, press the <ENTER> key.

After the two behavior and two phase variables are entered, the following command can be utilized to create a "data frame" that can be saved as a ".cvs" file for later use:

>**ssd<-data.frame(cry,pcry,esteem,pesteem)**

To view the data frame, shown in Figure A.1 (see p. 176), type the following command:

>**ssd**

Note how the "NA"s line up for each behavior variable and its corresponding phase variable. Also notice that each element is numbered on the far left with values ranging from the first ("1") to the last ("22").

Now the data frame can be saved as a *.csv* file by using the following function:

>**Savecsv()**

Remove the data frame from memory using the following command:

>**rm(list = ls())**

The data you just saved can be opened using the **Getcsv()** function and then **attach(ssd)**. It can also be opened and modified using Excel.

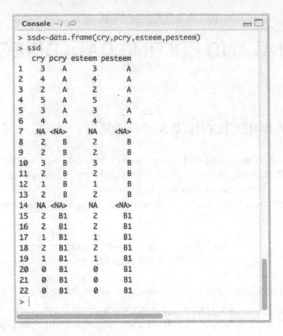

Figure A.1   The *ssd* data frame with four variables.

USING THE R DATA EDITOR

After opening a dataset using **Getcsv()** and using **attach(ssd)**, it can be modified using the following *R* command:

>**fix(ssd)**

When you do this, the "*R* Data Editor" will open, which allows you to make changes to your data in a spreadsheet-type format, as shown in Figure A.2 (see p. 177). After changing any data in the editor, remember to save it using the **Savecsv()** function.

IMPORTANT NOTE: This is an example of the Mac R Data Editor. The Windows version of the editor looks slightly different.

Use the arrow keys to move left and right or up and down the editor. A column can be added by moving the right arrow key past the last column, "pesteem." As shown in Figure A.3 (see p. 177) below, the column is labeled "var5."

Now double click on "var5" and a menu with attributes will appear. Select "Change name," as shown in Figure A.4 (see p. 178). Now change the name of the variable to *comment*.

Now enter "went home sick" in the comment column and third row. The results are shown in Figure A.5 (see p. 178) below.

Now type **ssd** in the Console to view your changes, displayed in Figure A.6 (see p. 178).

You can now use the **Savecsv()** function to save your changes.

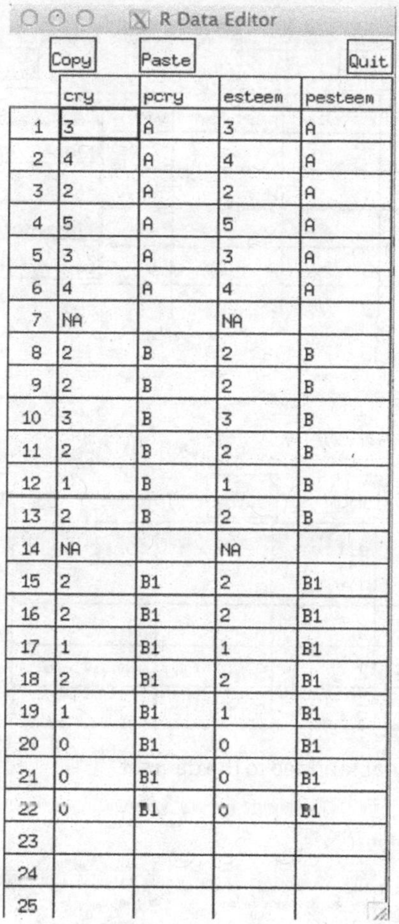

	cry	pcry	esteem	pesteem
1	3	A	3	A
2	4	A	4	A
3	2	A	2	A
4	5	A	5	A
5	3	A	3	A
6	4	A	4	A
7	NA		NA	
8	2	B	2	B
9	2	B	2	B
10	3	B	3	B
11	2	B	2	B
12	1	B	1	B
13	2	B	2	B
14	NA		NA	
15	2	B1	2	B1
16	2	B1	2	B1
17	1	B1	1	B1
18	2	B1	2	B1
19	1	B1	1	B1
20	0	B1	0	B1
21	0	B1	0	B1
22	0	B1	0	B1
23				
24				
25				

Figure A.2   The **ssd** data displayed in *R* Data Editor

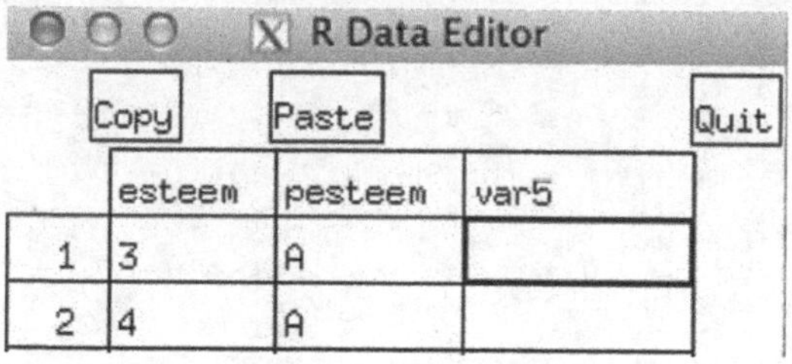

	esteem	pesteem	var5
1	3	A	
2	4	A	

Figure A.3   Adding a variable

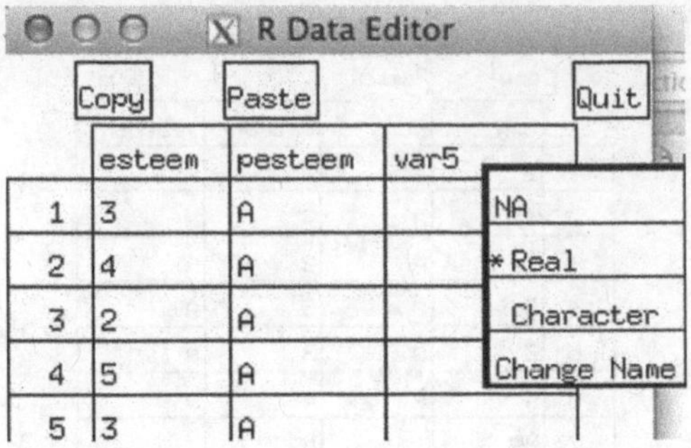

Figure A.4 Renaming a variable

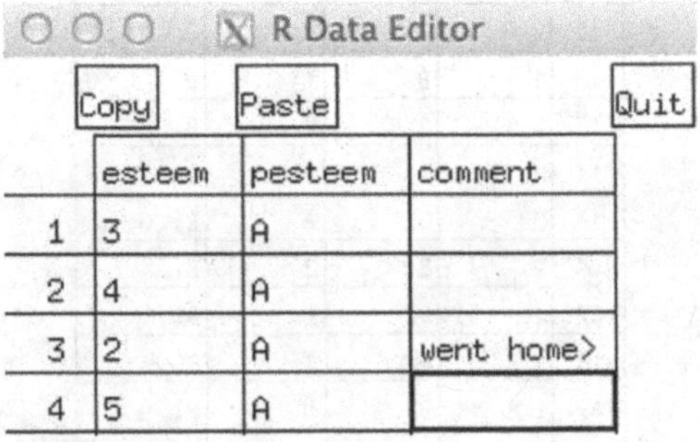

Figure A.5 The *comment* variable added to the data set

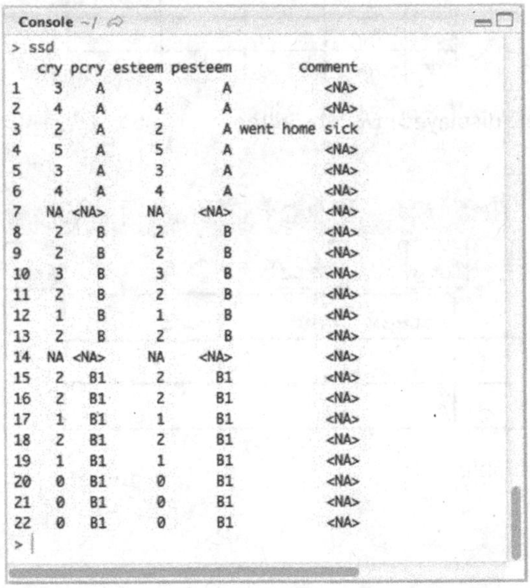

Figure A.6 The **ssd** with the added *comment* variable

SSD for R Quick Functions Guide

SSD for R Quick Functions Guide

Basic Functions

Open/Import.csv file

  Description: This function opens ".csv" file created in Excel.

  Command: **Getcsv()**

  Ex: **Getcsv()**

List all functions

  Description: List all functions available in *SSD for R*.

  Command: **SSDforR()**

  Ex: **SSDforR()**

List variables

  Description: Lists all variables in active data frame.

  Command: **listnames()**

  Ex: **listnames()**

Save a data file

  Description: Save a data file edited in *SSD for R* as ".csv" file.

  Command: **Savecsv()**

  Ex: **Savecsv()**

Graphing Functions

Create a line graph

  Description: This function builds a simple line chart for a given behavior across all phases.
    A space separates each phase.

  Command: **ABplot(*behavior, phaseX, ABxlab, ABylab, ABmain*)**

  behavior: behavior variable

  phaseX: phase variable

  ABxlab: label for *x*-axis; written between quotation marks

  ABylab: label for *y*-axis; written between quotation marks

  ABmain: main label for graph; written between quotation marks

  Ex: ABplot(yell, pyell, "school days", "yelling incidents", "Jenny's Yelling")

Create multiple line graphs in a single window

  Description: This function builds a simple line chart for a given behavior across all phases.
    A space separates each phase. This function needs to be invoked for each graph separately.
    plotnum() must precede use of this function.

  Command: **ABplotm(*behavior, phaseX, ABxlab, ABylab, ABmain*)**

    behavior: behavior variable

    phaseX: phase variable

    ABxlab: label for *x*-axis; written between quotation marks

    ABylab: label for *y*-axis; written between quotation marks

    ABmain: main label for graph; written between quotation marks

  Ex: **ABplotm(yell, pyell, "school days", "yelling incidents", "Jenny's Yelling")**

Sets the graphic environment when creating multiple line graphs
> *Description*: This function specifies the number of rows and columns to display in the graphics window when multiple line graphs are to be built.
> *Command*: **plotnum(*nr, nc*)**
>> nr: number of rows desired
>> nc: number of columns desired
> *Ex*: **plotnum(4,2)**

Draw a solid line between phases on a graph
> *Description*: This function enables the user to draw solid vertical lines between phases on a graph. Once the function is invoked, the user is prompted to accept the line or not.
> *Command*: **ABlines(behavior)**
>> behavior: behavior variable
> *Ex*: **ABlines(yell)**

Draw a dashed line between phases on a graph
> *Description*: This function enables the user to draw dashed vertical lines between phases on a graph. Once the function is invoked, the user is prompted to accept the line or not.
> *Command*: **ABlineD(*behavior*)**
>> behavior: behavior variable
> *Ex*: **ABlineD(yell)**

Label a graph
> *Description*: This function enables the user to write text on a graph. Users have three options for doing this: text with neither subscripts nor superscripts, text with superscripts, and text with subscripts. Commands for each and examples to label a graph are displayed below. After the command is invoked, users will be prompted to place the cursor where the text is to begin. After text is placed, users will be prompted to choose whether or not to accept the graph with the text.
> *Command*: **(for text with neither subscripts nor superscripts)**
> **ABtext()**
>> Text to be entered on graph must appear between quotation marks
> *Ex*: **ABtext("baseline")**
> *Command*: **(for text with superscripts) ABtext(expression(*text^superscript*))**
>> text: this is the text that is to be displayed on the graph
>> superscript: this is the actual superscript that is to be displayed
> *Ex*: **ABtext(expression(X^2))**
> *Command*: **(for text with subscripts) ABtext(expression(*text[subscript]*))**
>> text: this is the text that is to be displayed on the graph
>> subscript: this is the actual subscript that is to be displayed
> *Ex*: **ABtext(expression(B[1]))**

Draw an arrow on a graph
> *Description*: This function enables users to draw an arrow on a graph. For example, an arrow can be drawn from a text label of a critical event to a point on the graph. Once the function is invoked, the user is prompted to accept the arrow or not.
> *Command*: **ABarrow()**
> *Ex*: **ABarrow()**

Adding statistical lines (mean, median, or sd) to an **ABplot**
> *Description*: This function enables users to draw a line in a phase representing the mean, median, or standard deviation.
> *Command*: **ABstat(*behavior, phaseX, v, statX*)**
>> behavior: behavior variable

phaseX: phase variable

v: phase letter entered between quotes (e.g., "A", "B")

statX: mean, median, or sd written between quotation marks (e.g., "mean")

Ex: **ABstat(yell, pyell, "A", "mean")**

Adding trimmed mean line to an **ABplot**

Description: This function enables a user to add a line representing the trimmed mean to any phase of an **ABplot**.

Command: **Trimline(*behavior, phaseX, v*)**

behavior: behavior variable

phaseX: phase variable

v: phase letter entered between quotes (e.g., "A", "B")

Ex: **Trimline(yell,pyell,"B")**

Adding goal line to an **ABplot**

Description: This function enables a user to add a line representing the goal line (a level of function set as a goal for a client) to an **ABplot**.

Command: **Gline(*y*)**

y: y ordinate

Ex: **Gline(6)**

Adding standard deviation (SD) bands line an **ABplot**

Description: This function enables a user to add a line representing the SD bands (±1, ±2, or ±3) to any phase of an **ABplot**.

Command: **SDAband(*behavior, phaseX, v,bandX*)**

behavior: behavior variable

phaseX: phase variable

v: phase letter entered between quotes (e.g., "A", "B")

bandX: SDordinate 1, 2, or 3

Ex: **SDAband(yell,pyell,"A",2)**

Adding interquartile range (IQR) line to an **ABplot**.

Description: This function enables a user to add a line representing the interquartile range to any phase of an **ABplot**.

Command: **IQRline(*behavior, phaseX, v*)**

behavior: behavior variable

phaseX: phase variable

v: phase letter entered between quotes (e.g., "A", "B")

Ex: **IQRline(yell,pyell,"A")**

Create a ±1 standard deviation (sd) band graph for a given phase

Description: This function builds a ±1 standard deviation band graph for a given behavior based on a phase of the user's choice. A space separates each phase.

Command: **sd1bandgraph(*behavior, phaseX, v1, ABxlab, ABylab, ABmain*)**

behavior: behavior variable

phaseX: phase variable

v1: phase letter entered between quotes (e.g., "A", "B")

ABxlab: label for *x*-axis; written between quotation marks

ABylab: label for *y*-axis; written between quotation marks

ABmain: main label for graph; written between quotation marks

Ex: **sd1bandgraph(yell, pyell, "A", "school days", "yelling incidents", "SD graph: Jenny's Yelling")**

Create a ±1 standard deviation (sd) band graph across all phases
>*Description*: This function builds a ±1 standard deviation band graph for a given behavior across all phases. A space separates each phase.
>
>*Command*: **SD1(*behavior, phaseX, v1, ABxlab, ABylab, ABmain*)**
>>behavior: behavior variable
>>phaseX: phase variable
>>v1: phase letter entered between quotes (e.g., "A", "B")
>>ABxlab: label for *x*-axis; written between quotation marks
>>ABylab: label for *y*-axis; written between quotation marks
>>ABmain: main label for graph; written between quotation marks
>
>*Ex*: **SD1(yell, pyell, "A", "school days", "yelling incidents", "SD graph: Jenny's Yelling")**

Places legend at bottom of one standard deviation (SD1) band graph
>*Description*: This function enables the user to place a legend on a graph.
>NOTE: Once this legend is in place, the graph can no longer be altered.
>*Command*: **SD1legend()**
>*Ex*: **SD1legend()**

Create a ±2 standard deviation (sd) band graph for a given phase
>*Description*: This function builds a ±2 standard deviation band graph for a given behavior based on a phase of the user's choice. A space separates each phase.
>
>*Command*: **sd2bandgraph(*behavior, phaseX, v1,ABxlab, ABylab, ABmain*)**
>>behavior: behavior variable
>>phaseX: phase variable
>>v1: phase letter entered between quotes (e.g., "A", "B")
>>ABxlab: label for *x*-axis; written between quotation marks
>>ABylab: label for *y*-axis; written between quotation marks
>>ABmain: main label for graph; written between quotation marks
>
>*Ex*: **sd2bandgraph(yell, pyell, "A", "school days", "yelling incidents", "SD graph: Jenny's Yelling")**

Create a ±2 standard deviation (sd) band graph across all phases
>*Description*: This function builds a ±2 standard deviation band graph for a given behavior across all phases. A space separates each phase.
>
>*Command*: **SD2(*behavior, phaseX, v1,ABxlab, ABylab, ABmain*)**
>>behavior: behavior variable
>>phaseX: phase variable
>>v1: phase letter entered between quotes (e.g., "A", "B")
>>ABxlab: label for *x*-axis; written between quotation marks
>>ABylab: label for *y*-axis; written between quotation marks
>>ABmain: main label for graph; written between quotation marks
>
>*Ex*: **SD2(yell, pyell, "A", "school days", "yelling incidents", "SD graph: Jenny's Yelling")**

Places legend at bottom of two standard deviation (SD2) band graph
>*Description*: This function enables the user to place a legend on the graph.
>NOTE: Once this legend is in place, the graph can no longer be altered.
>*Command*: **SD2legend()**
>*Ex*: **SD2legend()**

Create an interquartile (iqr) band graph for a single phase
>*Description*: This function builds an iqr band graph for a given behavior based on a phase of the user's choice. The graph will only be drawn for the phase selected. Statistical output in the Console shows the interquartile bands.
>*Command*: **IQRbandgraph(*behavior, phaseX, v1,ABxlab, ABylab, ABmain*)**
>>behavior: behavior variable

phaseX: phase variable

v1: phase letter entered between quotes (e.g., "A", "B")

ABxlab: label for *x*-axis; written between quotation marks

ABylab: label for *y*-axis; written between quotation marks

ABmain: main label for graph; written between quotation marks

Ex: IQRbandgraph(yell, pyell, "A", "school days", "yelling incidents", "SD graph: Jenny's Yelling")

Create a interquartile (iqr) band graph through all phases

Description: This function builds an iqr band graph for a given behavior based on a phase of the user's choice. The graph will be drawn for all phases. A space separates phases. Statistical output in the Console shows the interquartile bands.

Command: **ABiqr(*behavior, phaseX, v1,ABxlab, ABylab, ABmain*)**

behavior: behavior variable

phaseX: phase variable

v1: phase label entered between quotes ("A","B","B1")

ABxlab: label for *x*-axis written between quotation marks

ABylab: label for *y*-axis written between quotation marks

ABmain: main label for graph between quotation marks

Ex: **ABiqr(cry, pcry, "A", "school days", "yelling incidents", "SD graph: Jenny's Crying")**

Places legend at bottom of **IQRbandgraph** or **ABiqr** band graph

Description: This function enables the user to place a legend on a graph. NOTE: Once this legend is in place, the graph can no longer be altered.

Command: **IQRlegend()**

Ex: **IQRlegend()**

Create a time series line graph

Description: This function builds a time series chart for a given behavior across all phases. A space separates each phase. There are no connecting dots.

Command: **ABtsplot(*behavior, phaseX, ABxlab, ABylab, ABmain*)**

behavior: behavior variable

phaseX: phase variable

ABxlab: label for *x*-axis; written between quotation marks

ABylab: label for *y*-axis; written between quotation marks

ABmain: main label for graph; written between quotation marks

Ex: **ABtsplot(yell, pyell, "school days", "yelling incidents", "Jenny's Yelling")**

Basic Statistical Analysis

Compute descriptive statistics for any phase

Description: This function produces descriptive statistics for all phases. Statistics produced are mean, 10% trimmed mean, median, standard deviation (sd), coefficient of variation (CV), range, interquartile range, and quantiles. Graphical output for this function is a box plot of data in each phase.

Command: **ABdescrip(*behavior, phaseX*)**

behavior: behavior variable

phaseX: phase variable on which test is based

Ex: **ABdescrip(yell, pyell)**

Ordinary least squares (OLS) regression for a single phase

Description: Conducts OLS regression for any phase. Coefficients and residuals are produced. Also a simple line graph for the specified phase with a regression line is displayed in the graph window.

Command: **Aregres (*behavior, phaseX, v1*)**

behavior: behavior variable

phaseX: phase variable on which test is based

v1: phase letter entered between quotes (e.g., "A", "B")

Ex: **Aregres(cry, pcry, "A")**

Robust regression for a single phase

Description: Conducts robust regression for any phase. Coefficients and residuals are produced. Also a simple line graph for the specified phase with a regression line is displayed in the graph window.

Command: **Arobust(***behavior, phaseX, v1***)**

behavior: behavior variable

phaseX: phase variable on which test is based

v1: phase letter entered between quotes (e.g., "A", "B")

Ex: **Arobust(cry, pcry, "A")**

OLS regression to compare phases

Description: Conducts OLS regression comparing any two phases. Coefficients and residuals are produced for each phase. Also a graph with a regression line is displayed for each phase in the graph window.

Command: **ABregres (***behavior, phaseX, v1, v2***)**

behavior: behavior variable

phaseX: phase variable on which test is based

v1: phase letter entered between quotes (e.g., "A", "B")

v2: phase letter entered between quotes (e.g., "A", "B")

Ex: **ABregres(cry, pcry, "A", "B")**

Robust regression to compare phases

Description: Conducts robust regression comparing any two phases. Coefficients and residuals are produced for each phase. Also a graph with a robust regression line is displayed for each phase in the graph window.

Command: **ABrobust(***behavior, phaseX, v1, v2***)**

behavior: behavior variable

phaseX: phase variable on which test is based

v1: phase letter entered between quotes (e.g., "A", "B")

v2: phase letter entered between quotes (e.g., "A", "B")

Ex: **ABrobust(cry, pcry, "A", "B")**

Effect Size Functions

Calculates most common effect size indices

Description: Displays the percentage change and calculated values for both the ES and d-index for any two phases. Information for interpreting calculated values appears in the Console.

Command: **Effectsize(***behavior, phaseX, v1, v2***)**

behavior: behavior variable

phaseX: phase variable on which test is based

v1: phase letter entered between quotes (e.g., "A", "B")

v2: phase letter entered between quotes (e.g., "A", "B")

Ex: **Effectsize(cry, pcry, "A", "B")**

Calculates g-index

Description: Calculates effect size based on scores in the desired zone.

Command: **Gindex(***behavior, phaseX, v1, v2***)**

behavior: behavior variable

phaseX: phase variable on which test is based

v1: phase letter entered between quotes (e.g., "A", "B")

v2: phase letter entered between quotes (e.g., "A", "B")
    Ex: **Gindex(cry, pcry, "A", "B")**

Calculation of Improvement Rate Difference
    Description: Calculates Improvement Rate Difference (IRD) and displays a graph. An increase in the targeted behavior is desired.
    Command: **IRDabove(***behavior, phaseX, v1, v2***)**
        behavior: behavior variable
        phaseX: phase variable on which test is based
        v1: phase letter entered between quotes (e.g., "A", "B")
        v2: phase letter entered between quotes (e.g., "A", "B")
    Ex: **IRDabove(cry, pcry,"A", "B")**

Calculation of Percentage of All Non-Overlapping Data.
    Description: This function evaluates the Percentage of All Non-overlapping Data (PAND).
        A decrease in the targeted behavior is desired.
    Command: **IRDbelow(***behavior, phaseX, v1, v2***)**
        behavior: behavior variable
        phaseX: phase variable on which test is based
        v1: phase letter entered between quotes (e.g., "A", "B")
        v2: phase letter entered between quotes (e.g., "A", "B")
    Ex: **IRDbelow(cry, pcry,"A", "B")**

Calculation of Percentage of All Non-Overlapping Data.
    Description: This function evaluates the Percentage of All Non-overlapping Data (PAND).
        An increase in the targeted behavior is desired.
    Command: **PANDabove(***behavior, phaseX, v1, v2***)**
        behavior: behavior variable
        phaseX: phase variable on which test is based
        v1: phase letter entered between quotes (e.g., "A", "B")
        v2: phase letter entered between quotes (e.g., "A", "B")
    Ex: **PANDabove(cry, pcry,"A", "B")**

Calculation of Percentage of All Non-overlapping Data.
    Description: This function evaluates the Percentage of All Non-overlapping Data (PAND).
        A decrease in the targeted behavior is desired.
    Command: **PANDbelow(***behavior, phaseX, v1, v2***)**
        behavior: behavior variable
        phaseX: phase variable on which test is based
        v1: phase letter entered between quotes (e.g., "A", "B")
        v2: phase letter entered between quotes (e.g., "A", "B")
    Ex: **PANDbelow(cry, pcry,"A", "B")**

Calculation of Percentage of All Non-overlapping Data.
    Description: This function calculates the Percentage of All Non-overlap of All Pairs (NAP).
        An increase in the targeted behavior is desired.
    Command: **NAPabove(***behavior, phaseX, v1, v2***)**
        behavior: behavior variable
        phaseX: phase variable on which test is based
        v1: phase letter entered between quotes (e.g., "A", "B")
        v2: phase letter entered between quotes (e.g., "A", "B")
    Ex: **NAPabove(cry, pcry,"A", "B")**

Calculation of Percentage of All Non-Overlapping Data

Description: This function calculates the Percentage of All Non-overlapping Pairs (NAP). A decrease in the targeted behavior is desired.

Command: **NAPbelow(***behavior, phaseX, v1, v2***)**

behavior: behavior variable

phaseX: phase variable on which test is based

v1: phase letter entered between quotes (e.g., "A", "B")

v2: phase letter entered between quotes (e.g., "A", "B")

Ex: **NAPbelow(cry, pcry,"A", "B")**

Calculation of Percentage of Data Exceeding the Median above the median

Description: The Percentage of Data Exceeding the Median (PEM) procedure offers a method to assess effect size and adjust for the influence of outliers in the baseline phase when desired values are above the reference line.

Command: **PEMabove(***behavior, phaseX, v1, v2***)**

behavior: behavior variable

phaseX: phase variable on which test is based

v1: phase letter entered between quotes (e.g., "A", "B")

v2: phase letter entered between quotes (e.g., "A", "B")

Ex: **PEMabove(cry, pcry,"A", "B")**

Calculation of Percentage of Data Exceeding the Median below the median

Description: The Percentage of Data Exceeding the Median (PEM) procedure offers a method to assess effect size and adjust for the influence of outliers in the baseline phase when desired values are below the reference line.

Command: **PEMbelow(***behavior, phaseX, v1, v2***)**

behavior: behavior variable

phaseX: phase variable on which test is based

v1: phase letter entered between quotes (e.g., "A", "B")

v2: phase letter entered between quotes (e.g., "A", "B")

Ex: **PEMbelow(cry, pcry,"A", "B")**

Create a legend on a PEM graph

Description: Adds a legend to a PEM graph. The graph cannot be modified in any way after the legend is added.

Command: **PEMlegend()**

Ex: **PEMlegend()**

Calculation of Percentage of Non-overlapping Data above the reference line

Description: The Percentage of Non-overlapping Data (PND) procedure offers a method to assess effect size based on the highest data point in the comparison phase.

Command: **PNDabove(***behavior, phaseX, v1, v2***)**

behavior: behavior variable

phaseX: phase variable on which test is based

v1: phase letter entered between quotes (e.g., "A", "B")

v2: phase letter entered between quotes (e.g., "A", "B")

Ex: **PNDabove(cry, pcry,"A", "B")**

Calculation of Percentage of Non-overlapping Data below the reference line

Description: The Percentage of Non-overlapping Data (PND) procedure offers a method to assess effect size based on the lowest data point in the comparison phase.

Command: **PNDbelow(***behavior, phaseX, v1, v2***)**

behavior: behavior variable

phaseX: phase variable on which test is based

v1: phase letter entered between quotes (e.g., "A", "B")

v2: phase letter entered between quotes (e.g., "A", "B")

Ex: **PNDbelow(cry, pcry,"A", "B")**

Create a legend on a PND graph

Description: Adds a legend to a PND graph. The graph cannot be modified in any way after the legend is added.

Command: **PNDlegend()**

Ex: **PNDlegend()**

Statistical Process Control (SPC) Charts

Create an X-bar range chart ($\overline{X}$-R chart)

Description: This function builds an $\overline{X}$-R chart using range. A space separates each phase. Used with multiple observations per sample. In the example below, there are seven observations (days) per sample (weeks). The "admitweek" is the grouping variable. If there were 8 weeks, the values for this variable would range from 1 to 8 with 7 observations for each. Values for the upper band (Uband), mean, and lower band (Lband) appear as statistical output in the Console.

Command: **XRchart(*behavior, groupX, bandX, ABxlab, ABylab, ABmain*)**

behavior: behavior variable

groupX: grouping variable

bandX: number of SDs desired (i.e., 1, 2, 3)

ABxlab: label for *x*-axis; written between quotation marks

ABylab: label for *y*-axis; written between quotation marks

ABmain: main label for graph; written between quotation marks

Ex: **XRchart(admits, admitweek, 3, "weeks","mean % of admits","Social Admits")**

Create an R chart using mean range

Description: This function builds an R chart using the mean range. A space separates each phase. Used with multiple observations per sample. In the example below, there are seven observations (days) per sample (weeks). The "admitweek" is the grouping variable. If there were 8 weeks, the values for this variable would range from 1 to 8 with 7 observations for each. This version of the R chart is recommended when the sample size is small (less than 10). It uses the mean range of the samples to track variation. Similar to the $\overline{X}$-R chart, using the mean range improves confidence in the measure. Values for the Uband, mean, and Lband appear as statistical output in the Console.

Command: **Rchart(*behavior, groupX, bandX, ABxlab, ABylab, ABmain*)**

behavior: behavior variable

groupX: grouping variable

bandX: number of SDs desired (i.e., 1, 2, 3)

ABxlab: label for *x*-axis; written between quotation marks

ABylab: label for *y*-axis; written between quotation marks

ABmain: main label for graph; written between quotation marks

Ex: **Rchart(admits, admitweek, 3,"weeks","mean range % of admits", "Social Admits")**

Create an R chart using standard deviation

Description: This function builds an R chart using standard deviation. A space separates each phase. Used with multiple observations per sample. In the example below, there are seven observations (days) per sample (weeks). The "admitweek" is the grouping variable. If there were 8 weeks, the values for this variable would range from 1 to 8 with 7 observations for each. This version of the R chart can be used with samples greater than 10. Values for the Uband, mean, and Lband appear as statistical output in the Console.

Command: **Rchartsd(*behavior, groupX, bandX, ABxlab, ABylab, ABmain*)**

behavior: behavior variable

groupX: grouping variable

bandX: number of SDs desired (i.e., 1, 2, 3)

ABxlab: label for *x*-axis; written between quotation marks

ABylab: label for *y*-axis; written between quotation marks

ABmain: main label for graph; written between quotation marks

Ex: **Rchartsd (admits, admitweek, 3, "weeks", "mean range % of admits", "Social Admits")**

Create line on R charts

Description: This function enables the user to draw solid vertical lines between phases on the SPC R chart using standard deviation and R chart using mean range. The user clicks the mouse on a upper and lower *y*-ordinate.

Command:**SPCline()**

Ex: **SPCline()**

Create a proportion chart (P chart)

Description: This function builds a p chart. A space separates each phase. Used with multiple observations per sample. The behavior variable must be binary. Values for the Uband, mean, and Lband appear as statistical output in the Console.

Command: **Pchart(*behavior, groupX, bandX, ABxlab, ABylab, ABmain*)**

behavior: behavior variable

groupX: grouping variable

bandX: number of SDs desired (i.e., 1, 2, 3)

ABxlab: label for *x*-axis; written between quotation marks

ABylab: label for *y*-axis; written between quotation marks

ABmain: main label for graph; written between quotation marks

Ex: **Pchart(group, wgroup, 3, "weeks", "proportion of attendance", "Jenny's Group Attendance")**

Create an X-moving range chart (X-mR chart)

Description: This function builds an X-mR chart and is used with individual data. A space separates each phase. The X-mR chart can be use to detect changes within and between phases. This chart should not be used when there is a trend in the data. As with the previous charts, large unexpected change in the undesired zone (values above upper band or below the lower band) may indicate the need to modify the intervention. Values for the Uband, mean, and Lband appear as statistical output in the Console.

Command: **Xmrchart(*behavior, phaseX, v1,bandX,ABxlab, ABylab, ABmain*)**

behavior: behavior variable

phaseX: phase variable on which bands are based

v1: phase letter entered between quotes (e.g., "A", "B")

bandX: number of SDs desired (i.e., 1, 2, 3)

ABxlab: label for *x*-axis; written between quotation marks

ABylab: label for *y*-axis; written between quotation marks

ABmain: main label for graph; written between quotation marks

Ex: **Xmrchart(esteem, pesteem, "A", 3,"weeks","mean self-esteem", "Jenny's Self-esteem")**

Create a C chart

Description: This function builds a C chart and is used with individual (i.e., ungrouped) data. A space separates each phase. For use when the outcome variable is a count (i.e., ratio level) variable.

Command: **Cchart(*behavior, phaseX, v1,bandX,ABxlab, ABylab, ABmain*)**

behavior: behavior variable

phaseX: phase variable on which bands are based

v1: phase letter entered between quotes (e.g., "A", "B")

bandX: number of SDs desired (i.e., 1, 2, 3)

ABxlab: label for *x*-axis; written between quotation marks

ABylab: label for *y*-axis; written between quotation marks

ABmain: main label for graph; written between quotation marks

Ex: **Cchart(yell, pyell, "A", 3, "days", "Count", "Count of Yelling")**

Places legend at bottom of any SPC band graph

Description: This function enables the user to place legend on a graph. NOTE: Once this legend is in place, the graph can no longer be altered.

Command: **SPClegend()**

Ex: **SPClegend()**

Tests and Functions Related to Autocorrelation

Test for lag-1 autocorrelation

Description: This function tests for lag-1 autocorrelation. This should be used any time the sample size is less than six. Any phase can be tested. Also produces a regression line graph.

Command: **ABrf2(*behavior, phaseX, v1*)**

behavior: behavior variable

phaseX: phase variable on which test is based

v1: phase letter entered between quotes (e.g., "A", "B")

Ex: **ABrf2(cry,pcry,"A")**

Test for autocorrelation for any lag

Description: This function tests for autocorrelation for any lag. Should be used with samples greater than or equal to six. Also produces a significance graph for lags. The Box-Ljung test of significance is performed for all lags up to and including the specified one.

Command: **ABautoacf (*behavior, phaseX, v, lags*)**

behavior: behavior variable

phaseX: phase variable on which test is based

v: phase letter entered between quotes (e.g., "A", "B")

lags: number of lags to be tested

Ex: **ABautoacf(cry, pcry,"A", 3)**

Test for partial autocorrelation for any lag

Description: This function tests for partial autocorrelation for any lag. Should only be used with samples greater than or equal to six. Also produces significance graph for lags.

Command: **ABautopacf (*behavior, phaseX, v, lags*)**

behavior: behavior variable

phaseX: phase variable on which test is based

v: phase letter entered between quotes (e.g., "A", "B")

lags: number of lags to be tested

Ex: **ABautopacf (cry, pcry,"A", 3)**

First difference transformation

Description: This function produces a first difference transformation for any phase. The results can be saved for later use by answering "y" to the prompt to save results. A line graph is produced for the original and transformed data. Statistical output includes displaying first differencing data in the Console.

Command: **diffchart(*behavior, phaseX, v1*)**

behavior: behavior variable

phaseX: phase variable on which test is based

v1: phase letter entered between quotes (e.g., "A", "B")

Ex: **diffchart(cry, pcry,"A")**

Moving average transformation

Description: This function produces a moving average transformation for any phase with every two scores being averaged. The results can be saved for later use by answering "y" to the prompt to save results. A line graph is produced for the original and transformed data. Statistical output includes displaying moving average data in the Console.

Command: **ABma (***behavior, phaseX, v1***)**

behavior: behavior variable

phaseX: phase variable on which test is based

v1: phase letter entered between quotes (e.g., "A", "B")

Ex: **ABma (cry, pcry, "A")**

Combine two or more data files

Description: This function combines data files. This is useful after data are created during transformations when using the **diffchart** or **ABma** functions. Once files with different phases are combined, you can use the saved file for significance testing.

Command: **Append()**

Ex: **Append()**

Tests of Statistical Significance

Proportion-frequency/binomial test

Description: Binomial test comparing the number of observations of a phase in a desired zone to another phase. User needs to select the method for defining a desired zone (e.g., below one sd)

Command: **ABbinomial(***phaseX, v1, v2, sucessA, sucessB***)**

phaseX: phase variable on which test is based

v1: phase letter entered between quotes (e.g., "A", "B")

v2: phase letter entered between quotes (e.g., "A", "B")

successA: number of observations in desired zone for v1

successB: number of observations in desired zone for v1

Ex: **ABbinomial(pyell,"A","B", 1, 15)**

Conservative dual-criteria test with desired zone above the lines

Description: Conservative dual-criteria test comparing the frequency of observations above both the mean and OLS regression line in any two phases.

Command: **CDCabove(***behavior, phaseX, v1, v2***)**

behavior: behavior variable

phaseX: phase variable on which test is based

v1: phase letter entered between quotes (e.g., "A", "B")

v2: phase letter entered between quotes (e.g., "A", "B")

Ex: **CDCabove(esteem, pesteem,"A","B")**

Conservative dual-criteria test with desired zone below the lines

Description: Conservative dual-criteria test comparing the frequency of observations below both the mean and OLS regression line in any two phases.

Command: **CDCbelow(***behavior, phaseX, v1, v2***)**

behavior: behavior variable

phaseX: phase variable on which test is based

v1: phase letter entered between quotes (e.g., "A", "B")

v2: phase letter entered between quotes (e.g., "A", "B")

Ex: **CDCbelow(esteem, pesteem,"A","B")**

Conservative dual-criteria test with desired zone above the lines using robust regression

Description: Conservative dual-criteria test comparing the frequency of observations above both the mean and robust regression line in any two phases.

Command: **RobustCDCabove(***behavior, phaseX, v1, v2***)**

behavior: behavior variable

phaseX: phase variable on which test is based

v1: phase letter entered between quotes (e.g., "A", "B")

v2: phase letter entered between quotes (e.g., "A", "B")
Ex: **RobustCDCabove(esteem, pesteem,"A","B")**

Conservative dual-criteria test with desired zone below the lines using robust regression
   Description: Conservative dual-criteria test comparing the frequency of observations below
      both the mean and robust regression line in any two phases.
   Command: **RobustCDCbelow(***behavior, phaseX, v1, v2***)**
      behavior: behavior variable
      phaseX: phase variable on which test is based
      v1: phase letter entered between quotes (e.g., "A", "B")
      v2: phase letter entered between quotes (e.g., "A", "B")
   Ex: **RobustCDCbelow(esteem, pesteem,"A","B")**

Chi-square test with desired zone above the mean
   Description: Chi-square test comparing the frequency of observations above the mean in any
      two phases.
   Command: **meanabove(***behavior, phaseX, v1, v2***)**
      behavior: behavior variable
      phaseX: phase variable on which test is based
      v1: phase letter entered between quotes (e.g., "A", "B")
      v2: phase letter entered between quotes (e.g., "A", "B")
   Ex: **meanabove(esteem, pesteem,"A","B")**

Chi-square test with desired zone below the mean
   Description: Chi-square test comparing the frequency of observations below the mean in any
      two phases.
   Command: **meanbelow(***behavior, phaseX, v1,v2***)**
      behavior: behavior variable
      phaseX: phase variable on which test is based
      v1: phase letter entered between quotes (e.g., "A", "B")
      v2: phase letter entered between quotes (e.g., "A", "B")
   Ex: **meanbelow(yell,pyell,"A","B")**

Chi-square test with desired zone above the median
   Description: Chi-square test comparing the frequency of observations above the median in
      any two phases.
   Command: **medabove(***behavior, phaseX, v1, v2***)**
      behavior: behavior variable
      phaseX: phase variable on which test is based
      v1: phase letter entered between quotes (e.g., "A", "B")
      v2: phase letter entered between quotes (e.g., "A", "B")
   Ex: **medabove(esteem, pesteem,"A","B")**

Chi-square with desired zone below the median
   Description: Chi-square test comparing the frequency of observations below the median in
      any two phases.
   Command: **medbelow(***behavior, phaseX,v1,v2***)**
      behavior: behavior variable
      phaseX: phase variable on which test is based
      v1: phase letter entered between quotes (e.g., "A", "B")
      v2: phase letter entered between quotes (e.g., "A", "B")
   Ex: **medbelow(yell,pyell,"A","B")**

Chi-square with desired zone above the trimmed mean

Description: Chi-square test comparing the frequency of observations above the trimmed mean in any two phases.

Command: **trimabove(***behavior, phaseX, v1, v2***)**

    behavior: behavior variable

    phaseX: phase variable on which test is based

    v1: phase letter entered between quotes (e.g., "A", "B")

    v2: phase letter entered between quotes (e.g., "A", "B")

Ex: **trimabove(esteem, pesteem,"A","B")**

Chi-square with desired zone below the trimmed mean

Description: Chi-square test comparing the frequency of observations below the trimmed mean in any two phases.

Command: **trimbelow(***behavior, phaseX,v1,v2***)**

    behavior: behavior variable

    phaseX: phase variable on which test is based

    v1: phase letter entered between quotes (e.g., "A", "B")

    v2: phase letter entered between quotes (e.g., "A", "B")

Ex: **trimbelow(yell,pyell,"A","B")**

Chi-square with desired zone above the OLS regression line

Description: Chi-square test comparing the frequency of observations above the OLS regression line in any two phases.

Command: **regabove(***behavior, phaseX, v1, v2***)**

    behavior: behavior variable

    phaseX: phase variable on which test is based

    v1: phase letter entered between quotes (e.g., "A", "B")

    v2: phase letter entered between quotes (e.g., "A", "B")

Ex: **regabove(esteem,pesteem,"A","B")**

Chi-square with desired zone below the OLS regression line

Description: Chi-square test comparing the frequency of observations below the OLS regression line in any two phases.

Command: **regbelow(***behavior, phaseX, v1, v2***)**

    behavior: behavior variable

    phaseX: phase variable on which test is based

    v1: phase letter entered between quotes (e.g., "A", "B")

    v2: phase letter entered between quotes (e.g., "A", "B")

Ex: **regbelow(cry,pcry,"A","B")**

Chi-square with desired zone above the robust regression line

Description: Chi-square test comparing the frequency of observations above the robust regression line in any two phases.

Command: **robregabove(***behavior, phaseX, v1, v2***)**

    behavior: behavior variable

    phaseX: phase variable on which test is based

    v1: phase letter entered between quotes (e.g., "A", "B")

    v2: phase letter entered between quotes (e.g., "A", "B")

Ex: **robregabove(esteem, pesteem,"A","B")**

Chi-square with desired zone below the robust regression line

Description: Chi-square test comparing the frequency of observations below the robust regression line in any two phases.

Command: **robregbelow(***behavior, phaseX, v1, v2***)**

    behavior: behavior variable

phaseX: phase variable on which test is based

v1: phase letter entered between quotes (e.g., "A", "B")

v2: phase letter entered between quotes (e.g., "A", "B")

Ex: **robregbelow(cry,pcry,"A","B")**

t test

Description: Computes Student *t* test between any two phases. This test should only be used if there is not a trend in either phase AND there is no problem with autocorrelation in either phase. Graphical output is a bar chart displaying the mean for each phase.

Command: **ABttest(*behavior, phaseX, v1,v2*)**

behavior: behavior variable

phaseX: phase variable on which test is based

v1: phase letter entered between quotes (e.g., "A", "B")

v2: phase letter entered between quotes (e.g., "A", "B")

Ex: **ABttest(yell, pyell,"A","B")**

Wilcoxon Rank Sum

Description: performs a two-sample Wilcoxon test between any two phases.

Nonparametric test to compare means. This test should only be used if there is not a trend in either phase AND there is no problem with autocorrelation in either phase.

Command: **ABWilcox(*behavior, phaseX, v1, v2*)**

behavior: behavior variable

phaseX: phase variable on which test is based

v1: phase letter entered between quotes (e.g., "A", "B")

v2: phase letter entered between quotes (e.g., "A", "B")

Ex: **ABWilcox(yell, pyell,"A","B")**

One-way analysis of variance

Description: Computes one-way ANOVA and performs Tukey multiple comparison test. This test should only be used if there is not a trend in any phase AND there is no problem with autocorrelation in any phase. Use ANOVA instead of a *t* test when comparing more than two phases

Command: **ABanova(*behavior, phaseX*)**

behavior: behavior variable

phaseX: phase variable on which test is based

Ex: **ABanova(esteem,pesteem)**

Scientific notation

Description: Coverts scientific notation to five decimal places

Command: **SN(*value*)**

value: value to be translated

Ex: **SN(2.73e-16)**

Group Data Functions

Lag-1 autocorrelation for group data

Description: This function tests for lag-1 autocorrelation for group data. Any phase can be tested. Produces a regression line graph

Command: **GABrf2(*behavior, phaseX, timeX, v1*)**

behavior: behavior variable

phaseX: phase variable on which test is based

timeX: time variable (e.g., week)

v1: letter of phase being tested (e.g., "A")

Ex: **GABrf2(attend, pattend, week, "A")**

t test for group data

> *Description*: Computes t test for group data. A bar graph showing the mean for each phase is displayed.
>
> *Command*: **GABttest(***behavior, phaseX, timeX, v1, v2***)**
>
> > behavior: behavior variable
> >
> > phaseX: phase variable on which test is based
> >
> > timeX: time variable (e.g., week)
> >
> > v1: letter of first phase (e.g., "A")
> >
> > v2: letter of second phase (e.g., "B")
>
> *Ex*: **GABttest(attend, pattend, week, "A", "B")**

Draws median line group data

> *Description*: Places median line for a single phase in group box plot.
>
> *Command*: **Gmedian(***behavior, phaseX, v***)**
>
> > behavior: behavior variable
> >
> > phaseX: phase variable on which test is based
> >
> > v: letter of phase (e.g., "A")
>
> *Ex*: **Gmedian(attend, pattend, "A")**

Functions Related to RMarkdown

Draw an arrow in an *RMarkdown* document

> *Description*: This function allows a user to draw an arrow on a *RMarkdown* graph. For example, an arrow can be drawn from a text label of a critical event to a data point on the graph.
>
> *Command*: **RMarrow(X1,Y1,X2,Y2)**
>
> *Ex*: **RMarrow(5.9,5,9.5,5.1)**

Draw a solid line between phases on a *RMarkdown* graph.

> *Description*: This function enables the user to draw solid vertical lines between phases on a graph created in *RMarkdown*.
>
> *Command*: **RMlines(behavior, X)**
>
> EX: **RMlines(yell,15.5)**

Adding goal line to an **ABplot** in an *RMarkdown* document.

> *Description*: This function enables a user to add a line representing the goal line (a level of function set as a goal for a client) to an **ABplot** in an *RMarkdown* document.
>
> *Command*: **RMGline(***y***)**
>
> > y: *y*-ordinate
>
> *Ex*: *RMGline(6)*
>
> *Command*: **RMstat(***behavior, phaseX, v, statX***)**
>
> > behavior: behavior variable
> >
> > phaseX: phase variable
> >
> > v: phase letter entered between quotes (e.g., "A", "B")
> >
> > X: start of horizontal line
> >
> > statX: mean, median, or sd written between quotation marks (e.g., "mean")
>
> *Ex*: **RMstat(yell, pyell, "A", "mean", 7)**

Adding trimmed mean line to an **ABplot**.

> *Description*: This function enables a user to add a line representing the trimmed mean to any phase of an **ABplot**.

Label an *RMarkdown* graph

> *Description*: This function enables the user to write text on a graph. Users have three options for doing this: text with neither subscripts nor superscripts, text with superscripts, and text with subscripts. Commands for each and examples to label a graph are displayed

below. After the command is invoked, users will be prompted to place the cursor where the text is to begin. After text is placed, users will be prompted to choose whether or not to accept the graph with the text.

Command: (for text with neither subscripts nor superscripts)
RMtext()
Text to be entered on graph must appear between quotation marks.
Ex: **RMtext("baseline",5,6)**
Command: (for text with superscripts) **ABtext(expression(*text^superscript,x,y*))**
text: This is the text that is to be displayed on the graph.
superscript: This is the actual superscript that is to be displayed.
x: *x* coordinate
y: *y* coordinate
Ex: **ABtext(expression(X^2,5,2))**
Command: (for text with subscripts) **ABtext(expression(*text[subscript]*,x,y))**
text: This is the text that is to be displayed on the graph.
subscript: This is the actual subscript that is to be displayed.
Ex: **ABtext(expression(B[1],8,2))**

Meta-analysis Functions

Calculation of mean effect size
Description: This function calculates a mean and SD for Cohen's D effect sizes. A file containing saved effect sizes must be opened by Getcsv() and then attached.
Command: **meanES(*ES, lab, esmain*)**
ES: effect size variable
lab: label for effect size variable
esmain: main title for graph
Ex: **meanES(ES,"cry","ES For Crying")**

Calculation of nonparametric mean effect size
Description: This function calculates a mean and SD for NAP effect sizes. A file containing saved NAP effect sizes must be opened by **Getcsv()** and then attached.
Command: **meanNAP *ES, lab, esmain*)**
ES: effect size variable
lab: label for effect size variable
esmain: main title for graph
Ex: **meanNAP(ES,"cry","ES For Crying")**

Meta Regression
Description: Meta regression for saved effect sizes. A file containing saved effect sizes must be opened by **Getcsv()** and then attached.
Command: **metareg(*ES, V*)**
ES: effect size variable
V: Variance of the effect size
Ex: **metareg(ES,V)**

Meta Regression with a moderator
Description: Meta regression with moderaor for saved effect sizes. A file containing saved effect sizes must be opened by **Getcsv()** and then attached.
Command: **metaregi(*ES, I, V*)**
ES: effect size variable
V: Variance of the effect size
Ex: **metaregi(ES,I,V)**

Interactive Command	*RMarkdown* Substitute	Syntax
ABarrow()	RMarrow()	RMarrow(X1,Y1,X2,Y2) X1 y1 **From** coordinates of arrow. X2 y2 **To** coordinates of arrow.
Gline()	RMGline()	RMGline(Y) **From** Y coordinate to draw horizontal line
ABlines()	RMlines()	RMlines(behavior, X) Behavior variable to draw vertical line **From** X coordinate to draw vertical line
ABstat	RMstat()	RMstat(behavior, phase,"statistic",X) X start of horizontal line
ABtext()	RMtext()	RMtext("text",X,Y) **From** X and Y coordinates to draw text
Getcsv()	ssd<-read()	ssd <- read.csv("file") file data file name

Decision Trees

SPC Chart Decision Tree

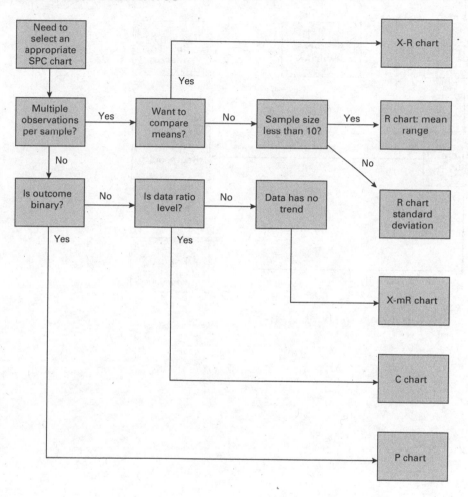

Chi-Square Decision Tree

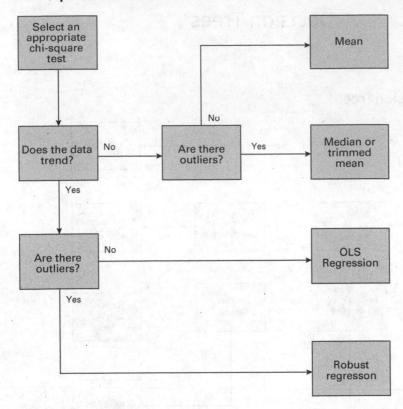

Hypothesis Test Decision Tree

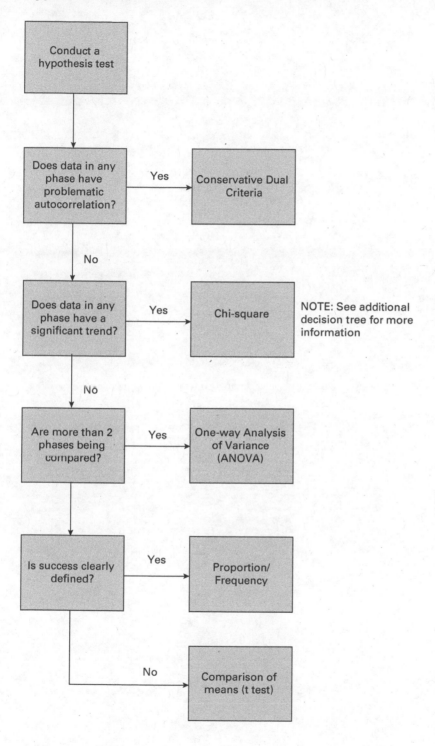

APPENDIX D

Bibliography of Additional Resources

Texts Covering Single-Subject Research Methodology

Bloom, M., Fischer, J., & Orme, J. G. (2009). *Evaluating practice: Guidelines for the accountable professional* (6th ed.). New York, NY: Pearson.

Franklin, R. D., Allison, D. B., & Gorman, B. S. (Eds.). (1997). *Design and analysis of single-case research*. Mahwah, NJ: Erlbaum.

Gast, D. L., & Ledford, J. (2010). *Single-subject research in behavioral sciences*. New York, NY: Routledge.

Golper, L. A. C. (2012). *Outcomes in speech-language pathology: Contemporary theories, models, and practices*. New York, NY: Thieme Medical.

Janosky, J. E., Leininger, S. L., Hoerger, M. P., & Libkuman, T. M. (2009). *Single subject designs in biomedicine* (2009 ed.). New York, NY: Springer.

Kazdin, A. E. (2011). *Single-case research designs: Methods for clinical and applied settings* (2nd ed.). New York, NY: Oxford University Press.

Kennedy, C. H. (2005). *Single-case designs for educational research*. New York, NY: Pearson Education.

Kratochwill, T. R., & Levin, J. R. (Eds.). (1992). *Single-case research design and analysis: New directions for psychology and education*. Hillsdale, NJ: Erlbaum.

Kratochwill, T. R., Hitchcock, J., Horner, R. H., Levin, J. R., Odom, S. L., Rindskopf, D. M. & Shadish, W. R. (2010). *Single-case designs technical documentation*. Retrieved from What Works Clearinghouse website: http://ies.ed.gov/ncee/wwc/pdf/wwc_scd.pdf

Nugent, W. R. (2009). *Analyzing single system design data*. New York, NY: Oxford University Press.

Orme, J. G., & Combs-Orme, T. (2011). *Outcome-informed evidence-based practice*. New York, NY: Pearson.

Ray, W. J. (2011). *Methods toward a science of behavior and experience* (10th ed.). Belmont, CA: Cengage Learning.

Richards, S., & Taylor, R. L. (2014). *Single subject research: Applications in educational and clinical settings* (2nd ed.). Belmont, CA: Wadsworth/Thomson Learning.

Riley-Tillman, T. C., & Burns, M. K. (2009). *Evaluating educational interventions: Single-case design for measuring response to intervention*. New York, NY: Guilford Press.

Rubin, A., & Babbie, E. R. (2013). *Research methods for social work* (8th ed.). Belmont, CA: Brooks/Cole.

Schweigert, W. A. (2011). *Research methods in psychology: A handbook* (3rd ed.). Long Grove, IL: Waveland Press.

Scruggs, T. E., & Mastropieri, M. A. (Eds.). (2006). *Applications of research methodology*. San Diego, CA: JAI Press.

Shadish, W. R., Hedges, L. V., & Pustejovsky, J. E. (2014). Analysis and meta-analysis of single-case designs with a standardized mean difference statistic: A primer and applications. *Journal of School Psychology, 52*(2), 123–147.

Skinner, C. H. (2013). *Single-subject designs for school psychologists*. New York, NY: Routledge.

Thomas, J., & Hersen, M. (2011). *Understanding research in clinical and counseling psychology* (2nd ed.). New York, NY: Routledge.

Thyer, B. A. (2009). *The handbook of social work research methods*. Thousand Oaks, CA: Sage.

Thyer, B. A., & Myers, L. L. (2007). *A social worker's guide to evaluating practice outcomes*. Alexandria, VA: CSWE Press.

Vannest, K. J., Davis, J. L., and Parker, R. I. (2013). *A new approach to single case research*. New York, NY: Routledge.

Wallen, N. E., & Fraenkel, J. R. (2013). *Educational research: A guide to the process* (2nd ed.). New York, NY: Routledge.

Recommended Reading on SPC Charts

Mitra, A. (2008). Control charts for attributes. In *Fundamentals of quality control and improvement* (3rd ed., pp. 369–414). Hoboken, NJ: Wiley.

Orme, J. G., & Cox, M. E. (2001). Analyzing single-subject design data using statistical process control charts. *Social Work Research, 25*(2), 115–127.

Recommended Reading Rmarkdown

Baumer, B., & Udwin, D. (2015). R markdown. Wiley Interdisciplinary Reviews: Computational Statistics, *7*(3), 167–177.

Wickham, H., & Grolemund, G. (2016). *R for data science: Import, tidy, transform, visualize, and model data.* O'Reilly Media.

Xie, Y., Allaire, J. J., & Grolemund, G. (2018). *R Markdown: The definitive guide.* CRC Press.

Xie, Y., Dervieux, C., & Riederer, E. (2020). *R Markdown cookbook.* CRC Press.

Helpful Websites

Campbell Collaboration—http://www.campbellcollaboration.org. Produces freely available systematic reviews in a number of broad disciplines, including crime and justice, education, international development, and social welfare. This site contains many free resources, including training on methodology related to the production of systematic reviews.

Gitlab. https://evoldyn.gitlab.io/evomics-2018/ref-sheets/rmarkdown-cheatsheet-2.0.pdf. Cheat sheet for RMarkdown.

Outcome Informed Evidence-Based Practice. http://ormebook.com/. This site accompanies Orme and Combs-Orme's (2011) text listed in the Texts Covering Single-Subject Research Methodology. In addition to information relevant to each chapter in the text, this site contains a lot of pertinent information related to single-subject research in general, including bibliographies, templates for collecting data and creating graphs, and links to standardized scales.

R for Data Science. https://r4ds.had.co.nz. Contains an online free-to-use copy of the R *for Data Science* book.

RMarkdown. The definitive guide provides an online free-to-use version of R Markdown. https://bookdown.org/yihui/rmarkdown/

RStudio RMarkdown. https://rmarkdown.rstudio.com. Contains a host of recourses on using RMarkdown, including cheat sheets and a reference guide.

Single-Case Research. http://www.singlecaseresearch.org/. Information on single case research design and analysis through the posting of published papers, manuscripts in press, and white papers. Free calculators are available for the purpose of analysis along with instructional videos on a variety of analysis topics.

U.S. Department of Education's Institute of Education Sciences. http://ies.ed.gov/. Contains a host of resources related to education, including methodological papers and presentations, datasets, and access to the What Works Clearinghouse.

Additional Resource

Volkov, B. B., & King, J. A. (2007). A checklist for building organizational evaluation capacity. Retrieved from https://wmich.edu/sites/default/files/attachments/u350/2014/organiziationevalcapacity.pdf

References

Archer, B., Azios, J. H., Müller, N., & Macatangay, L. (2019). Effect sizes in single-case aphasia studies: A comparative, autocorrelation-oriented analysis. *Journal of Speech, Language, and Hearing Research, 62*(7), 2473–2482.

Arnold, M. E. (2006). Developing evaluation capacity in extension 4-H field faculty. A framework for success. *American Journal of Evaluation, 27*(2), 257–269.

Auerbach, C., & Schudrich, W. Z. (2013). SSD for R: A comprehensive statistical package to analyze single-system data. *Research on Social Work Practice, 23*(3), 346–353. doi:10.1177/1049731513477213

Auerbach, C., & Zeitlin , W. (2021). *SSD for R (version 1.5.2).* Vienna, Austria: R Foundation for Statistical Computing. Retrieved from http://www.R-project.org/

Baizerman, M., Compton, D. W., & Hueftle Stockdill, S. (2002). New directions for ECB. In D. W. Compton, M. Baizerman, & S. Stockdill (Eds.), *The art, craft, and science of evaluation capacity building* (Vol. 2002, pp. 109–120). San Francisco, CA: Jossey-Bass.

Baumer, B., & Udwin, D. (2015). R markdown. *Wiley Interdisciplinary Reviews: Computational Statistics, 7*(3), 167–177.

Beck, A. T., Ward, C. H., Mendelson, M., Mock, J., & Erbaugh, J. (1961). An inventory for measuring depression. *Archives of General Psychiatry, 4*, 561–571.

Beddoe, L. (2011). Investing in the future: Social workers talk about research. *British Journal of Social Work, 41*(3), 557–575.

Benneyan, J. C., Lloyd, R. C., & Plsek, P. E. (2003). Statistical process control as a tool for research and healthcare improvement. *Quality and Safety in Health Care, 12*(6), 458–464. doi:10.1136/qhc.12.6.458

Berger, R. (2010). EBP practitioners in search of evidence. *Journal of Social Work, 10*(2), 175–191.

Bloom, M., Fischer, J., & Orme, J. G. (2009). *Evaluating practice: Guidelines for the accountable professional* (6th ed.). New York, NY: Pearson.

Borckardt, J. J. (2008). User's guide: Simulation modeling analysis: Time series analysis program for short time series data streams: Version 8.3.3. Retrieved from http://www.clinicalresearcher.org/SMA_Guide.pdf

Brossart, D. F., Parker, R. I., & Castillo, L. G. (2011). Robust regression for single-case data analysis: How can it help? *Behavior Research Methods, 43*(3), 710–719. doi:10.3758/s13428-011-0079-7

Buck, J. A. (2011). The looming expansion and transformation of public substance abuse treatment under the Affordable Care Act. *Health Affairs, 30*(8), 1402–1410.

Carman, J. G., & Fredericks, K. A. (2010). Evaluation Capacity and Nonprofit Organizations: Is the glass half-empty or half-full? *American Journal of Evaluation, 31*(1), 84–104.

Chambless, D. L., Baker, M. J., Baucom, D. H., Beutler, L. E., Calhoun, K. S., Crits-Christoph, P., . . . Haaga, D. A. (1998). Update on empirically validated therapies, II. *Clinical Psychologist, 51*(1), 3–16.

Choi, M., & Ruona, W. E. (2011). Individual readiness for organizational change and its implications for human resource and organization development. *Human Resource Development Review, 10*(1), 46–73.

Christie, D., & Menter, I. (2009). Research capacity building in teacher education: Scottish collaborative approaches. *Journal of Education for Teaching, 35*(4), 337–354.

Cohen, J. (1988). *Statistical power analysis for the behavioral sciences* (2nd ed.). Hillsdale, NJ: Erlbaum.

Council on Social Work Education (CSWE). (2015). *Educational policy and accreditation standards.* Alexandria, VA: Author.

Edwards, J. R., & Cable, D. M. (2009). The value of value congruence. *Journal of Applied Psychology, 94*(3), 654–677.

Epstein, I., & Blumenfield, S. (2012). *Clinical data-mining in practice-based research: Social work in hospital settings.* Routledge.

Ferguson, C. J. (2009). An effect size primer: A guide for clinicians and researchers. *Professional Psychology: Research and Practice, 40*(5), 532–538.

Fisher, W. W., Kelley, M. E., & Lomas, J. E. (2003). Visual aids and structured criteria for improving visual inspection and interpretation of single-case designs. *Journal of Applied Behavior Analysis, 36*(3), 387–406.

Fouché, C., & Lunt, N. (2010). Nested mentoring relationships reflections on a practice project for mentoring research capacity amongst social work practitioners. *Journal of Social Work, 10*(4), 391–406.

Free Software Foundation, Inc. (2012). *RStudio*. Boston, MA: Author.

Gast, D. L., & Ledford, J. (2010). *Single-subject research in behavioral sciences*. New York, NY: Routledge.

Glass, G. V., McGaw, B., & Smith, M. L. (1981). *Meta-analysis in social research*. Thousand Oaks, CA: Sage.

Higgins, J., & Thomas, J. (Eds.). (2020). *Cochrane handbook for systematic reviews of interventions* (Version 6.1.0). Chichester, UK: Wiley. Retrieved from https://training.cochrane.org/handbook/current

Huitema, B. E., & McKean, J. W. (1994). Two reduced-bias autocorrelation estimators: rF1 and rF2. *Perceptual and Motor Skills, 78*(1), 323–330.

Janosky, J. E., Leininger, S. L., Hoerger, M. P., & Libkuman, T. M. (2009). *Single subject designs in biomedicine* (2009 ed.). New York, NY: Springer.

Kazdin, A. E. (2011). *Single-case research designs: Methods for clinical and applied settings* (2nd ed.). New York, NY: Oxford University Press.

Kratochwill, T. R., & Levin, J. R. (2014). *Single-case intervention research: Methodological and statistical advances*. American Psychological Association.

Kratochwill, T. R., Hitchcock, J. H., Horner, R. H., Levin, J. R., Odom, S. L., Rindskopf, D. M., & Shadish, W. R. (2013). Single-case intervention research design standards. *Remedial and Special Education, 34*(1), 26–38. doi:10.1177/0741932512452794

Kratochwill, T. R., Hitchcock, J., Horner, R. H., Levin, J. R., Odom, S. L., Rindskopf, D. M., & Shadish, W. R. (2010). *Single-case designs technical documentation*. Washington, DC: US Department of Education, Institute of Education Sciences. Retrieved from https://files.eric.ed.gov/fulltext/ED510743.pdf

Krishef, C. H. (1991). *Fundamental approaches to single subject design and analysis*. Malabar, FL: Krieger.

Kromrey, J. D., & Foster-Johnson, L. (1996). Determining the efficacy of intervention: The use of effect sizes for data analysis in single-subject research. *Journal of Experimental Education, 65*(1), 73–93.

Lenz, A. S. (2012). Calculating effect size in single-case research: A comparison of nonoverlap methods. *Measurement and Evaluation in Counseling and Development, 46*(1), 64–73. doi:10.1177/0748175612456401

Logan, L. R., Hickman, R. R., Harris, S. R., & Heriza, C. B. (2008). Single-subject research design: Recommendations for levels of evidence and quality rating. *Developmental Medicine & Child Neurology, 50*(2), 99–103.

Lunt, N., Fouché, C., & Yates, D. (2008). *Growing research in practice (GRIP): An innovative partnership model*. Wellington, New Zealand: Families Commission.

Ma, H.-H. (2009). The effectiveness of intervention on the behavior of individuals with autism. A meta-analysis using percentage of data points exceeding the median of baseline phase (PEM). *Behavior Modification, 33*(3), 339–359.

Macgowan, M. J. (2008). *A guide to evidence-based group work*. Oxford University Press.

Macgowan, M. J. (2012). A standards-based inventory of foundation competencies in social work with groups. *Research on Social Work Practice, 22*(5), 578–589. doi:10.1177/1049731512443288

Manolov, R., Solanas, A., Sierra, V., & Evans, J. J. (2011). Choosing among techniques for quantifying single-case intervention effectiveness. *Behavior Therapy, 42*(3), 533–545.

Matyas, T., & Greenwood, K. (1990). Visual analysis of single-case time series: Effects of variability, serial dependence, and magnitude of intervention effects. *Journal of Applied Behavioral Analysis, 23*(3), 341–351.

McCrystal, P., & Wilson, G. (2009). Research training and professional social work education: Developing research-minded practice. *Social Work Education, 28*(8), 856–872.

Mechanic, D. (2012). Seizing opportunities under the Affordable Care Act for transforming the mental and behavioral health system. *Health Affairs, 31*(2), 376–382.

Miller, B. (n.d.). *Single-subject research design (SSRD)*. Vancouver, BC, Canada: University of British Columbia School of Rehab Sciences.

Mitra, A. (2008). Control charts for attributes. In *Fundamentals of quality control and improvement* (3rd ed., pp. 369–414). Hoboken, NJ: Wiley.

Mohammed, M. A., & Worthington, P. (2012). Why traditional statistical process control charts for attribute data should be viewed alongside an XMR-chart. *BMJ Quality & Safety, 22*(3), 263–269. doi:10.1136/bmjqs-2012-001324

Morgan, D. L. (2008). *Single-case research methods for the behavioral and health sciences*. Sage.

Nagler, E., Rindskopf, D. M., & Shadish, W. R. (2008). *Analyzing data from small N designs using multilevel models: A procedural handbook.* New York: Graduate Center, CUNY.

Nathan, P. E., & Gorman, J. M. (Eds.). (2002). *A guide to treatments that work* (2nd ed.). New York, NY: Oxford University Press.

National Association of Social Workers. (2017). *Code of ethics*. Washington, DC.

Nourbakhsh, M. R., & Ottenbacher, K. J. (1994). The statistical analysis of single-subject data: A comparative examination. *Physical Therapy, 74*(8), 768–776.

Orme, J., & Powell, J. (2008). Building research capacity in social work: process and issues. *British Journal of Social Work, 38*(5), 988–1008.

Orme, J. G. (1991). Statistical conclusion validity for single-system designs. *Social Service Review, 65*(3), 468–491. doi:10.2307/30012410

Orme, J. G., & Combs-Orme, T. (2011). *Outcome-informed evidence-based practice*. New York: Pearson.

Orme, J. G., & Cox, M. E. (2001). Analyzing single-subject design data using statistical process control charts. *Social Work Research, 25*(2), 115–127.

Parker, R. I. (2006). Increased reliability for single-case research results: Is the bootstrap the answer? *Behavior Therapy, 37*(4), 326–338. doi:10.1016/j.beth.2006.01.007

Parker, R. I., Hagan-Burke, S., & Vannest, K. (2007). Percentage of all non-overlapping data (PAND). An alternative to PND. *Journal of Special Education, 40*(4), 194–204. doi:10.1177/00224669070400040101

Parker, R. I., Vannest, K. J., & Brown, L. (2009). The improvement rate difference for single-case research. *Exceptional Children, 75*(2), 135–150.

Parker, R. I., Vannest, K. J., & Davis, J. L. (2011). Effect size in single-case research: A review of nine nonoverlap techniques. *Behavior Modification, 35*(4), 303–322.

Polit, D. F., & Chaboyer, W. (2012). Statistical process control in nursing research. *Research in Nursing & Health, 35*(1), 82–93. doi:10.1002/nur.20467

Portney, L. G., & Watkins, M. P. (2008). *Foundations of clinical research: Applications to practice* (3rd ed.). Hoboken, NJ: Prentice Hall.

Preskill, H., & Boyle, S. (2008). A multidisciplinary model of evaluation capacity building. *American Journal of Evaluation, 29*(4), 443–459.

Pustejovsky, J. E. (2019). Procedural sensitivities of effect sizes for single-case designs with directly observed behavioral outcome measures. *Psychological Methods, 24*(2), 217.

R Core Team. (2013). *R: A language and environment for statistical computing*. Vienna, Austria: R Foundation for Statistical Computing. Retrieved from https://www.scirp.org/(S(351jmbntvnsjt1aadkposzje))/reference/ReferencesPapers.aspx?ReferenceID=1787696

Rock, B. D., Auerbach, C., Kaminsky, P., & Goldstein, M. (1993). Integration of computer and social work culture: A developmental model. In B. Glastonbury (Ed.), *Human welfare and technology: Papers from the Husita 3 Conference on IT and the quality of life and services*. Maastricht, Netherlands: Van Gorcum, Assen.

The R Project for Statistical Computing. (n.d.). *What is R?* Retrieved from http://www.r-project.org/about.html

Schlosser, R., & Wendt, O. (2008). Systematic reviews and meta-analyses of single-subject experimental designs (SSEDs). National Center for the Dissemination of Disability Research. https://ktdrr.org/systematicregistry/lib_systematic_search.cgi?location=sr&sel_1=104

Schudrich, W. (2012). Implementing a modified version of Parent Management Training (PMT) with an intellectually disabled client in a special education setting. *Journal of Evidence-Based Social Work, 9*(5), 421–423.

Scruggs, T. E., & Mastropieri, M. A. (1998). Summarizing single-subject research issues and applications. *Behavior Modification, 22*(3), 221–242. doi:10.1177/01454455980223001

Scruggs, T. E., & Mastropieri, M. A. (2013). PND at 25: Past, present, and future trends in summarizing single-subject research. *Remedial and Special Education, 34*(1), 9–19. doi:10.1177/0741932512440730

Shaw, I. (2005). Practitioner research: evidence or critique? *British Journal of Social Work, 35*(8), 1231–1248.

Shaw, I. (2011). *Evaluating in practice* (2nd ed.). Burlington, VT: Ashgate.

Smith, I. R., Garlick, B., Gardner, M. A., Brighouse, R. D., Foster, K. A., & Rivers, J. T. (2013). Use of graphical statistical process control tools to monitor and improve outcomes in cardiac surgery. *Heart, Lung and Circulation, 22*(2), 92–99. doi:10.1016/j.hlc.2012.08.060

Smith, J. D. (2012). Single-case experimental designs: A systematic review of published research and current standards. *Psychological Methods, 17*, 510–550.

Stewart, K. K., Carr, J. E., Brandt, C. W., & McHenry, M. M. (2007). An evaluation of the conservative dual-criterion method for teaching university students to visually inspect Ab-design graphs. *Journal of Applied Behavior Analysis, 40*(4), 713–718. doi:10.1901/jaba.2007.713-718

Swoboda, C. M., Kratochwill, T. R., & Levin, J. R. (2010). Conservative dual-criterion method for single-case research: A guide for visual analysis of AB, ABAB, and multiple-baseline designs. Wisconsin Center for Education Research Working Paper No. 2010-13. Retrieved from https://wcer.wisc.edu/docs/working-papers/Working_Paper_No_2010_13.pdf

Tasdemir, A. (2012). Effect of autocorrelation on the process control charts in monitoring of a coal washing plant. *Physicochemical Problems of Mineral Processing, 48*(2), 495–512.

Tate, R. L., & Perdices, M. (2019). *Single-case experimental designs for clinical research and neurorehabilitation settings: Planning, conduct, analysis and reporting.* Routledge.

Thombs, B. D., Ziegelstein, R. C., Beck, C. A., & Pilote, L. (2008). A general factor model for the Beck Depression Inventory-II: Validation in a sample of patients hospitalized with acute myocardial infarction. *Journal of Psychosomatic Research, 65*(2), 115–121.

Thyer, B. A., & Myers, L. L. (2011). The quest for evidence-based practice: A view from the United States. *Journal of Social Work, 11*(1), 8–25.

Vannest, K. J., Davis, J. L., & Parker, R. I. (2013). *A new approach to single case research.* New York: Routledge.

Verzani, J. (2004). *Using R for introductory statistics.* Chapman and Hall/CRC.

Volkov, B. B., & King, J. A. (2007). A checklist for building organizational evaluation capacity. Retrieved from https://wmich.edu/sites/default/files/attachments/u350/2014/organiziationevalcapacity.pdf

Weisberg, S., & Fox, J. (2010). *An R companion to applied regression* (2nd ed.). Thousand Oaks, CA: Sage.

Wendt, O. (2009, May). *Calculating effect sizes for single-subject experimental designs: An overview and comparison.* Presented at the Ninth Annual Campbell Collaboration Colloquium, Oslo, Norway.

Wendt, O., & Rindskopf, D. (2020). *Exploring new directions in statistical analysis of single-case experimental designs.*

Wheeler, D. J. (2004). *Advanced topics in statistical process control: The power of Shewhart's charts* (2nd ed.). Knoxville, TN: SPC Press.

Woodall, W. H. (2006). The use of control charts in health-care and public-health surveillance. *Journal of Quality Technology, 38*(2), 89–104.

Index